Lecture Notes in Computer Science 10519

Commenced Publication in 1973
Founding and Former Series Editors:
Gerhard Goos, Juris Hartmanis, and Jan van Leeuwen

More information about this series at http://www.springer.com/series/7407

Anthony Bonato · Fan Chung Graham
Paweł Prałat (Eds.)

Algorithms and Models for the Web Graph

14th International Workshop, WAW 2017
Toronto, ON, Canada, June 15–16, 2017
Revised Selected Papers

 Springer

Editors
Anthony Bonato
Ryerson University
Toronto, ON
Canada

Fan Chung Graham
Mathematics
UC San Diego
La Jolla, CA
USA

Paweł Prałat
Mathematics
Ryerson University
Toronto, ON
Canada

ISSN 0302-9743   ISSN 1611-3349 (electronic)
Lecture Notes in Computer Science
ISBN 978-3-319-67809-2   ISBN 978-3-319-67810-8 (eBook)
DOI 10.1007/978-3-319-67810-8

Library of Congress Control Number: 2017953396

LNCS Sublibrary: SL1 – Theoretical Computer Science and General Issues

This Springer imprint is published by Springer Nature
The registered company is Springer International Publishing AG
The registered company address is: Gewerbestrasse 11, 6330 Cham, Switzerland

Preface

The 14th Workshop on Algorithms and Models for the Web Graph (WAW 2017) took place at the Fields Institute for Research in Mathematical Sciences, Canada, June 15–16, 2017. This is an annual meeting, which is traditionally co-located with another, related, conference. WAW 2017 was co-located with the Canadian Discrete and Algorithmic Mathematics Conference (CanaDAM 2017) and was part of the Focus Program on Random Graphs and Applications to Complex Networks at the Fields Institute. Co-location of the workshop and conference provided opportunities for researchers in two different but interrelated areas to interact and to exchange research ideas. It was an effective venue for the dissemination of new results and for fostering research collaboration.

The World Wide Web has become part of our everyday life, and information retrieval and data mining on the Web are now of enormous practical interest. The algorithms supporting these activities combine the view of the Web as a text repository and as a graph, induced in various ways by links among pages, hosts, and users. The aim of the workshop was to further the understanding of graphs that arise from the Web and various user activities on the Web, and stimulate the development of high-performance algorithms and applications that exploit these graphs. The workshop gathered the researchers who are working on graph-theoretic and algorithmic aspects of related complex networks, including social networks, citation networks, biological networks, molecular networks, and other networks arising from the Internet.

This volume contains the papers presented during the workshop. Each submission was reviewed by Program Committee members. Papers were submitted and reviewed using the EasyChair online system. The committee members decided to accept seven papers.

June 2017

Anthony Bonato
Fan Chung Graham
Paweł Prałat

Organization

General Chairs

Andrei Z. Broder   Google Research, USA
Fan Chung Graham   University of California San Diego, USA

Organizing Committee

Anthony Bonato   Ryerson University, Canada
Fan Chung Graham   University of California, San Diego, USA
Paweł Prałat   Ryerson University, Canada

Sponsoring Institutions

The Fields Institute for Research in Mathematical Sciences, Canada
Natural Sciences and Engineering Research Council of Canada, Canada
Microsoft Research New England, USA
Google, USA
Moscow Institute of Physics and Technology, Russia
Yandex, Russia
Internet Mathematics, USA
Ryerson University, Canada

Program Committee

Konstantin Avratchenkov   Inria, France
Paolo Boldi   University of Milan, Italy
Anthony Bonato   Ryerson University, Canada
Milan Bradonjic   Bell Laboratories
Michael Brautbar   Toast, Inc.
Fan Chung Graham   UC San Diego, USA
Collin Cooper   King's College London, UK
Andrzej Dudek   Western Michigan University, USA
Alan Frieze   Carnegie Mellon University, USA
Aristides Gionis   Aalto University, Finland
David Gleich   Purdue University, USA
Jeannette Janssen   Dalhousie University, Canada
Julia Komjathy   Eindhoven University of Technology, The Netherlands
Ravi Kumar   Google
Silvio Lattanzi   Google

Marc Lelarge	Inria, France
Stefano Leonardi	Sapienza University of Rome, Italy
Nelly Litvak	University of Twente, The Netherlands
Michael Mahoney	UC Berkeley, USA
Oliver Mason	NUI Maynooth, Ireland
Dieter Mitsche	Université de Nice Sophia-Antipolis, France
Peter Morters	University of Bath, UK
Tobias Mueller	Utrecht University, The Netherlands
Liudmila Ostroumova	Yandex
Pan Peng	TU Dortmund, Germany
Xavier Perez-Gimenez	University of Nebraska-Lincoln, USA
Pawel Pralat	Ryerson University, Canada
Yana Volkovich	AppNexus
Stephen Young	Pacific Northwest National Laboratory, USA

Contents

Moment-Based Parameter Estimation in Binomial Random Intersection Graph Models

Joona Karjalainen[✉] and Lasse Leskelä

Aalto University, Espoo, Finland
`lasse.leskela@aalto.fi`
`http://math.aalto.fi/en/people/joona.karjalainen,`
`http://math.aalto.fi/~lleskela/`

Abstract. Binomial random intersection graphs can be used as parsimonious statistical models of large and sparse networks, with one parameter for the average degree and another for transitivity, the tendency of neighbours of a node to be connected. This paper discusses the estimation of these parameters from a single observed instance of the graph, using moment estimators based on observed degrees and frequencies of 2-stars and triangles. The observed data set is assumed to be a subgraph induced by a set of n_0 nodes sampled from the full set of n nodes. We prove the consistency of the proposed estimators by showing that the relative estimation error is small with high probability for $n_0 \gg n^{2/3} \gg 1$. As a byproduct, our analysis confirms that the empirical transitivity coefficient of the graph is with high probability close to the theoretical clustering coefficient of the model.

Keywords: Statistical network model · Network motif · Model fitting · Moment estimator · Sparse graph · Two-mode network · Overlapping communities

1 Introduction

Random intersection graphs are statistical network models with overlapping communities. In general, an intersection graph on a set of n nodes is defined by assigning each node i a set of attributes V_i, and then connecting those node pairs $\{i, j\}$ for which the intersection $V_i \cap V_j$ is nonempty. When the assignment of attributes is random we obtain a random undirected graph. By construction, this graph has a natural tendency to contain strongly connected communities because any set of nodes $W_k = \{i : V_i \ni k\}$ affiliated with attribute k forms a clique.

The simplest nontrivial model is the binomial random intersection graph $G = G(n, m, p)$ introduced in [13], having n nodes and m attributes, where any particular attribute k is assigned to a node i with probability p, independently of other node–attribute pairs. A statistical model of a large and sparse network with nontrivial clustering properties is obtained when n is large, $m \sim \beta n$ and $p \sim \gamma n^{-1}$ for

A. Bonato et al. (Eds.): WAW 2017, LNCS 10519, pp. 1–15, 2017.
DOI: 10.1007/978-3-319-67810-8_1

some constants β and γ. In this case the limiting model can be parameterised by its mean degree $\lambda = \beta\gamma^2$ and attribute intensity $\mu = \beta\gamma$. By extending the model by introducing random node weights, we obtain a statistical network model which is rich enough to admit heavy tails and nontrivial clustering properties [4,5,8,10]. Such models can also be generalised to the directed case [6]. An important feature of this class of models is the analytical tractability related to component sizes [3,15] and percolation dynamics [1,7].

In this paper we discuss the estimation of the model parameters based on a single observed instance of a subgraph induced by a set of n_0 nodes. We introduce moment estimators for λ and μ based on observed frequencies of 2-stars and triangles, and describe how these can be computed in time proportional to the product of the maximum degree and the number of observed nodes. We also prove that the statistical network model under study has a nontrivial empirical transitivity coefficient which can be approximated by a simple parametric formula in terms of μ.

The majority of classical literature on the statistical estimation of network models concerns exponential random graph models [19], whereas most of the recent works are focused on stochastic block models [2] and stochastic Kronecker graphs [9]. For binomial random intersection graphs with $m \ll n$, it has been shown [17] that the underlying attribute assignment can in principle be learned using maximum likelihood estimation. To the best of our knowledge, the current paper appears to be the first of its kind to discuss parameter estimation in random intersection graphs where m is of the same order as n.

The rest of the paper is organised as follows. In Sect. 2 we describe the model and its key assumptions. Section 3 summarises the main results. Section 4 describes numerical simulation experiments for the performance of the estimators. The proofs of the main results are given in Sect. 5, and Sect. 6 concludes the paper.

2   Model Description

2.1   Binomial Random Intersection Graph

The object of study is an undirected random graph $G = G(n, m, p)$ on node set $\{1, 2, \ldots, n\}$ with adjacency matrix having diagonal entries $A(i, i) = 0$ and off-diagonal entries

$$A(i, j) = \min\left(\sum_{k=1}^{m} B(i, k)B(j, k), \ 1\right),$$

where $B(i, k)$ are independent $\{0, 1\}$-valued random integers with mean p, indexed by $i = 1, \ldots, n$ and $k = 1, \ldots, m$. The matrix B represents a random assignment of m attributes to n nodes, both labeled using positive integers, so that $B(i, k) = 1$ when attribute k is assigned to node i. The set of attributes assigned to node i is denoted by $V_i = \{k : B(i, k) = 1\}$. Then a node pair $\{i, j\}$ is connected in G if and only if the intersection $V_i \cap V_j$ is nonempty.

2.2   Sparse and Balanced Parameter Regimes

We obtain a large and sparse random graph model by considering a sequence of graphs $G(n, m, p)$ with parameters $(n, m, p) = (n_\nu, m_\nu, p_\nu)$ indexed by a scale parameter $\nu \in \{1, 2, \dots\}$ such[1] that $n \gg 1$ and $p \ll m^{-1/2}$ as $\nu \to \infty$. In this case a pair of nodes $\{i, j\}$ is connected with probability

$$\mathbb{P}(ij \in E(G)) = 1 - (1 - p^2)^m \sim mp^2,$$

and the expected degree of a node i is given by

$$\mathbb{E} \deg_G(i) = (n - 1)\mathbb{P}(ij \in E(G)) \sim nmp^2. \qquad (2.1)$$

Especially, we obtain a large random graph with a finite limiting mean degree $\lambda \in (0, \infty)$ when we assume that

$$n \gg 1, \qquad mp^2 \sim \lambda n^{-1}. \qquad (2.2)$$

This will be called the *sparse parameter regime* with mean degree λ.

The most interesting model with nontrivial clustering properties is obtained when we also assume that $p \sim \mu m^{-1}$ for some constant $\mu \in (0, \infty)$. In this case the full set of conditions is equivalent to

$$n \gg 1, \qquad m \sim (\mu^2/\lambda)n, \qquad p \sim (\lambda/\mu)n^{-1}, \qquad (2.3)$$

and will be called as *balanced sparse parameter regime* with mean degree λ and attribute intensity μ.

2.3   Induced Subgraph Sampling

Assume that we have observed the subgraph $G^{(n_0)}$ of G induced by a set $V^{(n_0)}$ of n_0 nodes sampled from the full set of n nodes, so that $E(G^{(n_0)})$ consists of node pairs $\{i, j\} \in E(G)$ such that $i \in V^{(n_0)}$ and $j \in V^{(n_0)}$. The sampling mechanism used to generate $V^{(n_0)}$ is assumed to be stochastically independent of G. Especially, any nonrandom selection of $V^{(n_0)}$ fits this framework. On the other hand, several other natural sampling mechanisms [14] are ruled out by this assumption, although we believe that several of the results in this paper can be generalised to a wider context.

In what follows, we shall assume that the size of observed subgraph satisfies $n^\alpha \ll n_0 \leq n$ for some $\alpha \in (0, 1)$. An important special case with $n_0 = n$ amounts to observing the full graph G.

---

[1] For number sequences $f = f_\nu$ and $g = g_\nu$ indexed by integers $\nu \geq 1$, we denote $f \sim g$ if $f_\nu/g_\nu \to 1$ and $f \ll g$ if $f_\nu/g_\nu \to 0$ as $\nu \to \infty$. The scale parameter is usually omitted.

3   Main Results

3.1   Estimation of Mean Degree

Consider a random intersection graph $G = G(n, m, p)$ in a sparse parameter regime (2.2) with mean degree $\lambda \in (0, \infty)$, and assume that we have observed a subgraph $G^{(n_0)}$ of G induced by a set of nodes $V^{(n_0)}$ of size n_0, as described in Sect. 2.3. Then a natural estimator of λ is the normalised average degree

$$\hat{\lambda}(G^{(n_0)}) = \frac{n}{n_0^2} \sum_{i \in V^{(n_0)}} \deg_{G^{(n_0)}}(i). \tag{3.1}$$

This estimator is asymptotically unbiased because by (2.1),

$$\mathbb{E}\hat{\lambda}(G^{(n_0)}) = \frac{n}{n_0}(n_0 - 1)\mathbb{P}(ij \in E(G)) \sim \lambda.$$

The following result provides a sufficient condition for the consistency of the estimator of the mean degree λ, i.e., $\hat{\lambda} \to \lambda$ in probability as $n \to \infty$.

Theorem 3.1. *For a random intersection graph $G = G(n, m, p)$ in a sparse parameter regime (2.2), the estimator of λ defined by (3.1) is consistent when $n_0 \gg n^{1/2}$. Moreover, $\hat{\lambda}(G^{(n_0)}) = \lambda + O_p(n^{1/2}/n_0)$ for $m \gg n_0^2/n \gg 1$.*

3.2   Transitivity Coefficient

For a random or nonrandom graph G with maximum degree at least two, the transitivity coefficient (a.k.a. global clustering coefficient [12,16]) is defined by

$$t(G) = 3\frac{N_{K_3}(G)}{N_{S_2}(G)} \tag{3.2}$$

and the *model transitivity coefficient* by

$$\tau(G) = 3\frac{\mathbb{E}N_{K_3}(G)}{\mathbb{E}N_{S_2}(G)},$$

where $N_{K_3}(G)$ is the number of triangles[2] and $N_{S_2}(G)$ is the number of 2-stars[3] in G. The above definitions are motivated by noting that

$$t(G) = \mathbb{P}_G(I_2I_3 \in E(G) \mid I_1I_2 \in E(G), I_1I_3 \in E(G)),$$
$$\tau(G) = \mathbb{P}(I_2I_3 \in E(G) \mid I_1I_2 \in E(G), I_1I_3 \in E(G)),$$

for an ordered 3-tuple of distinct nodes (I_1, I_2, I_3) selected uniformly at random and independently of G, where $\mathbb{P}_G$ refers to conditional probability given

---

[2] subgraphs isomorphic to the graph K_3 with $V(K_3) = \{1, 2, 3\}$ and $E(K_3) = \{12, 13, 23\}$.

[3] subgraphs isomorphic to the graph S_2 with $V(S_2) = \{1, 2, 3\}$ and $E(S_2) = \{12, 13\}$.

an observed realisation of G. The model transitivity coefficient $\tau(G)$ is a non-random quantity which depends on the random graph model G only via its probability distribution, and is often easier to analyse than its empirical counterpart. Although $\tau(G) \neq \mathbb{E}t(G)$ in general, it is widely believed that $\tau(G)$ is a good approximation of $t(G)$ in large and sparse graphs [4,8]. The following result confirms this in the context of binomial random intersection graphs.

Theorem 3.2. *Consider a random intersection graph* $G = G(n, m, p)$ *in a balanced sparse parameter regime* (2.3). *If* $n_0 \gg n^{2/3}$, *then*

$$t(G^{(n_0)}) \;=\; \frac{1}{1+\mu} + o_p(1). \tag{3.3}$$

It has been observed (with a slightly different parameterisation) in [8] that the model transitivity coefficient of the random intersection graph $G = G(n, m, p)$ satisfies

$$\tau(G) \;=\; \begin{cases} 1 + o(1), & p \ll m^{-1}, \\ \frac{1}{1+\mu} + o(1), & p \sim \mu m^{-1}, \\ o(1), & m^{-1} \ll p \ll m^{-1/2}, \end{cases}$$

and only depends on n via the scale parameter. Hence, as a consequence of Theorem 3.2, it follows that

$$t(G) \;=\; \tau(G) + o_p(1)$$

for large random intersection graphs $G = G(n, m, p)$ in the balanced sparse parameter regime (2.3).

3.3   Estimation of Attribute Intensity

Consider a random intersection graph $G = G(n, m, p)$ in a balanced sparse parameter regime (2.3) with mean degree $\lambda \in (0, \infty)$ and attribute intensity $\mu \in (0, \infty)$, and assume that we have observed a subgraph $G^{(n_0)}$ of G induced by a set of nodes $V^{(n_0)}$ of size n_0, as described in Sect. 2.3. We will now introduce two estimators for the attribute intensity μ.

The first estimator of μ is motivated by the connection between the empirical and model transitivity coefficients established in Theorem 3.2. By ignoring the error term in (3.3), plugging the observed subgraph $G^{(n_0)}$ into the definition of the transitivity coefficient (3.2), and solving for μ, we obtain an estimator

$$\hat{\mu}_1(G^{(n_0)}) \;=\; \frac{N_{S_2}(G^{(n_0)})}{3 N_{K_3}(G^{(n_0)})} - 1. \tag{3.4}$$

An alternative estimator of μ is given by

$$\hat{\mu}_2(G^{(n_0)}) \;=\; \left(\frac{n_0 N_{S_2}(G^{(n_0)})}{2 N_{K_2}(G^{(n_0)})^2} - 1 \right)^{-1}, \tag{3.5}$$

where $N_{K_2}(G^{(n_0)}) = |E(G^{(n_0)})|$. A heuristic derivation of the above formula is as follows. For a random intersection graph G in the balanced sparse parameter regime (2.3), the expected number of 2-stars in $G^{(n_0)}$ is asymptotically (see Sect. 5)

$$\mathbb{E}N_{S_2}(G^{(n_0)}) \sim 3\binom{n_0}{3}(mp^3 + m^2p^4) \sim \frac{1}{2}n_0^3\mu^3(1+\mu)m^{-2}$$

and the expectation of $N_{K_2}(G^{(n_0)}) = |E(G^{(n_0)})|$ is asymptotically

$$\mathbb{E}N_{K_2}(G^{(n_0)}) \sim \binom{n_0}{2}mp^2 \sim \frac{1}{2}n_0^2\mu^2m^{-1}.$$

Hence

$$\frac{\mathbb{E}N_{S_2}(G^{(n_0)})}{(\mathbb{E}N_{K_2}(G^{(n_0)}))^2} \sim \frac{2}{n_0}(1+\mu^{-1}),$$

so by omitting the expectations above and solving for μ we obtain (3.5).

The following result confirms that both of the above heuristic derivations yield consistent estimators for the attribute intensity when the observed subgraph is large enough.

Theorem 3.3. *For a random intersection graph $G = G(n, m, p)$ in a balanced sparse parameter regime (2.3), the estimators of μ defined by (3.4) and (3.5) are consistent when $n_0 \gg n^{2/3}$.*

3.4   Computational Complexity of the Estimators

The evaluation of the estimator $\hat{\lambda}$ given by (3.1) requires computing the degrees of the nodes in the observed subgraph $G^{(n_0)}$. This can be done in $O(n_0 d_{\max})$ time, where $d_{\max}$ denotes the maximum degree of $G^{(n_0)}$.

Evaluating the estimator $\hat{\mu}_1$ given by (3.4) requires counting the number of triangles in $G^{(n_0)}$ which is a nontrivial task for very large graphs. A naive algorithm requires an overwhelming $O(n_0^3)$ time for this, a listing method can accomplish this in $O(n_0 d_{\max}^2)$ time, and there also exist various more advanced algorithms [18].

The estimator $\hat{\mu}_2$ given by (3.5) can be computed without the need to compute the number of triangles. Actually, the computation of $\hat{\mu}_2$ only requires to evaluate the degrees of the nodes in $G^{(n_0)}$. Namely, with help of the formulas

$$N_{K_2}(G^{(n_0)}) = \frac{1}{2}\sum_{i \in V^{(n_0)}} \deg_{G^{(n_0)}}(i) \quad \text{and} \quad N_{S_2}(G^{(n_0)}) = \sum_{i \in V^{(n_0)}} \binom{\deg_{G^{(n_0)}}(i)}{2},$$

one can verify that

$$\hat{\mu}_2(G^{(n_0)}) = \left(\frac{a_2 - a_1}{a_1^2} - 1\right)^{-1},$$

where $a_k = n_0^{-1} \sum_{i \in V^{(n_0)}} \deg_{G^{(n_0)}}(i)^k$ denotes the k-th moment of the empirical degree distribution of $G^{(n_0)}$.

We conclude that the parameters (λ, μ) of the random intersection graph $G = G(n, m, p)$ in the balanced sparse parameter regime (2.3) can be consistently estimated in $O(n_0 d_{\max})$ time using the estimators $\hat{\lambda}$ and $\hat{\mu}_2$.

4 Numerical Experiments

In this section we study the non-asymptotic behaviour of the parameter estimators $\hat{\lambda}$ (3.1), $\hat{\mu}_1$ (3.4), and $\hat{\mu}_2$ (3.5) using simulated data. In the first experiment, a random intersection graph was generated for each $n = 50, 70, \ldots, 1000$, using parameter values ($\lambda = 9$, $\mu = 3$) and ($\lambda = 2$, $\mu = 0.5$). All of the data was used for estimation, i.e., $n_0 = n$.

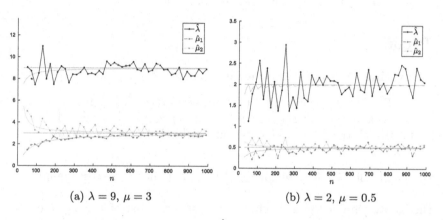

(a) $\lambda = 9$, $\mu = 3$ $\qquad\qquad$ (b) $\lambda = 2$, $\mu = 0.5$

Fig. 1. Simulated values of the estimators $\hat{\lambda}$, $\hat{\mu}_1$, and $\hat{\mu}_2$ with $n_0 = n$. The solid curves show the theoretical values of the estimators when the feature counts $N_*(G^{(n)})$ are replaced by their expected values.

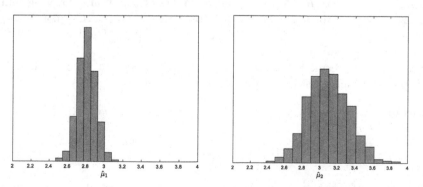

Fig. 2. 1000 simulated values of $\hat{\mu}_1$ and $\hat{\mu}_2$ with $\lambda = 9$, $\mu = 3$, and $n_0 = n = 750$.

Figure 1 shows the computed estimates $\hat{\lambda}$, $\hat{\mu}_1$, and $\hat{\mu}_2$ for each n. For comparison, the theoretical values of these estimators are also shown when the counts of links, 2-stars, and triangles are replaced by their expected values in (3.1), (3.4), and (3.5).

With ($\lambda = 9$, $\mu = 3$), the parameter μ is generally underestimated by $\hat{\mu}_1$ and overestimated by $\hat{\mu}_2$. The errors in $\hat{\mu}_1$ appear to be dominated by the bias, whereas the errors in $\hat{\mu}_2$ are mostly due to variance. With ($\lambda = 2$, $\mu = 0.5$), the simulated graphs are morse sparse. The differences between the two estimators of μ are small, and the relative error of $\hat{\lambda}$ appears to have increased. The discontinuities of the theoretical values of $\hat{\lambda}$ are due to the rounding of the numbers of attributes m.

In the second experiment, 1000 random intersection graphs were simulated with $n_0 = n = 750$ and ($\lambda = 9$, $\mu = 3$). Histograms of the estimates of μ are shown in Fig. 2. The bias is visible in both $\hat{\mu}_1$ and $\hat{\mu}_2$, and the variance of $\hat{\mu}_2$ is larger than that of $\hat{\mu}_1$. However, the difference in accuracy is counterbalanced by the fact that $\hat{\mu}_1$ requires counting the triangles.

5 Proofs

5.1 Covering Densities of Subgraphs

Denote by $\mathrm{Pow}(\Omega)$ the collection of all subsets of Ω. For $\mathcal{A}, \mathcal{B} \subset \mathrm{Pow}(\Omega)$ we denote $\mathcal{A} \Subset \mathcal{B}$ and say that $\mathcal{B}$ is a *covering family* of $\mathcal{A}$, if for every $A \in \mathcal{A}$ there exists $B \in \mathcal{B}$ such that $A \subset B$. A covering family $\mathcal{B}$ of $\mathcal{A}$ is called *minimal* if for any $B \in \mathcal{B}$,

(i) the family obtained by removing B from $\mathcal{B}$ is not a covering family of $\mathcal{A}$, and

(ii) the family obtained by replacing B by a strict subset of B is not a covering family of $\mathcal{A}$.

For a graph $R = (V(R), E(R))$, we denote by $\mathrm{MCF}(R)$ the set of minimal covering families of $E(R)$. Note that all members of a minimal covering family have size at least two. For a family of subsets $\mathcal{C} = \{C_1, \ldots, C_t\}$ consisting of t distinct sets, we denote $|\mathcal{C}| = t$ and $\|\mathcal{C}\| = \sum_{s=1}^{t} |C_s|$. The notation $R \subset G$ means that R is a subgraph of G.

The following result is similar in spirit to [13, Theorem 3], but focused on subgraph frequencies instead of appearance thresholds.

Theorem 5.1. *If $mp^2 \ll 1$, then for any finite graph R not depending on the scale parameter,*

$$\mathbb{P}(G \supset R) \sim \sum_{\mathcal{C} \in \mathrm{MCF}(R)} m^{|\mathcal{C}|} p^{\|\mathcal{C}\|}$$

The proof of Theorem 5.1 is based on two auxiliary results which are presented first.

Lemma 5.2. *For any intersection graph G on $\{1, \ldots, n\}$ generated by attribute sets $\mathcal{V} = \{V_1, \ldots, V_n\}$ and any graph R with $V(R) \subset V(G)$, the following are equivalent:*

(i) $R \subset G$.
(ii) $E(R) \Subset \mathcal{V}$.
(iii) There exists a family $\mathcal{C} \in \mathrm{MCF}(R)$ such that $E(R) \Subset \mathcal{C} \Subset \mathcal{V}$.

Proof. (i) $\Longleftrightarrow$ (ii). Observe that a node pair $e \in \binom{V}{2}$ satisfies $e \in E(G)$ if and only if $e \subset V_j$ for some $V_j \in \mathcal{V}$. Hence $E(R) \subset E(G)$ if and only if for every $e \in E(R)$ there exists $V_j \in \mathcal{V}$ such that $e \subset V_j$, or equivalently, $E(R) \Subset \mathcal{V}$.

(ii) $\Longrightarrow$ (iii). If $E(R) \Subset \mathcal{V}$, define $C_j = V_j \cap V(R)$. Then $\mathcal{C} = \{C_1, \ldots, C_m\}$ is a covering family of $E(R)$. Then test whether $\mathcal{C}$ still remains a covering family of $E(R)$ if one its members is removed. If yes, remove the member of $\mathcal{C}$ with the highest label. Repeat this procedure until we obtain a covering family $\mathcal{C}'$ of $E(R)$ for which no member can be removed. Then test whether some $C \in \mathcal{C}'$ can be replaced by a strict subset of C. If yes, do this replacement, and repeat this procedure until we obtain a covering family $\mathcal{C}''$ of $E(R)$ for which no member can be shrunk in this way. This mechanism implies that $\mathcal{C}''$ is a minimal covering family of $E(R)$, for which $E(R) \Subset \mathcal{C}'' \Subset \mathcal{V}$.

(iii) $\Longrightarrow$ (ii). Follows immediately from the transitivity of $\Subset$.

Lemma 5.3. *If $mp^2 \ll 1$, then for any scale-independent finite collection $\mathcal{C} = \{C_1, \ldots, C_t\}$ of finite subsets of $\{1, 2, \ldots\}$ of size at least 2, the probability that the family of attribute sets $\mathcal{V} = \{V_1, \ldots, V_n\}$ of $G = G(n, m, p)$ is a covering family of $\mathcal{C}$ satisfies*

$$\mathbb{P}(\mathcal{V} \Supset \mathcal{C}) \sim m^{|\mathcal{C}|} p^{||\mathcal{C}||}.$$

Proof. For $s = 1, \ldots, t$, denote by $N_s = \sum_{j=1}^m 1(V_j \supset C_s)$ the number of attribute sets covering C_s. Note that N_s follows a binomial distribution with parameters m and $p^{|C_s|}$. Because $|C_s| \geq 2$, it follows that the mean of N_s satisfies $mp^{|C_s|} \leq mp^2 \ll 1$. Using elementary computations related to the binomial distribution (see e.g. [13, Lemmas 1,2]) it follows that the random integers $N_1, \ldots, N_t$ are asymptotically independent with $\mathbb{P}(N_s \geq 1) \sim mp^{|C_s|}$, so that

$$\mathbb{P}(\mathcal{V} \Supset \mathcal{C}) = \mathbb{P}(N_1 \geq 1, \ldots, N_t \geq 1) \sim \prod_{s=1}^t \mathbb{P}(N_s \geq 1) \sim m^t p^{\sum_{s=1}^t |C_s|}.$$

Proof (Proof of Theorem 5.1). By Lemma 5.2, we see that

$$\mathbb{P}(G \supset R) = \mathbb{P}\left(\bigcup_{\mathcal{C} \in \mathrm{MCF}(R)} \{\mathcal{V} \Supset \mathcal{C}\} \right).$$

Bonferroni's inequalities hence imply $U_1 - U_2 \leq \mathbb{P}(G \supset R) \leq U_1$, where

$$U_1 = \sum_{\mathcal{C} \in \mathrm{MCF}(R)} \mathbb{P}(\mathcal{V} \Supset \mathcal{C}) \quad \text{and} \quad U_2 = \sum_{\mathcal{C}, \mathcal{D}} \mathbb{P}(\mathcal{V} \Supset \mathcal{C}, \mathcal{V} \Supset \mathcal{D}),$$

and the latter sum is taken over all unordered pairs of distinct minimal covering families $\mathcal{C}, \mathcal{D} \in \text{MCF}(R)$. Note that by Lemma 5.3,

$$U_1 \sim \sum_{\mathcal{C} \in \text{MCF}(R)} m^{|\mathcal{C}|} p^{\|\mathcal{C}\|},$$

so to complete the proof it suffices to verify that $U_2 \ll U_1$.

Fix some minimal covering families $\mathcal{C} = \{C_1, \ldots, C_s\}$ and $\mathcal{D} = \{D_1, \ldots, D_t\}$ of $E(R)$ such that $\mathcal{C} \neq \mathcal{D}$. Then either $\mathcal{C}$ has a member such that $C_i \notin \mathcal{D}$, or $\mathcal{D}$ has a member such that $D_j \notin \mathcal{C}$. In the former case $\mathcal{C} \cup \mathcal{D} \supset \{C_i, D_1, \ldots, D_t\}$, so that by Lemma 5.3,

$$\mathbb{P}(\mathcal{V} \ni \mathcal{C} \cup \mathcal{D}) \leq \mathbb{P}(\mathcal{V} \ni \{C_i, D_1, \ldots, D_t\}) \sim m^{t+1} p^{|C_i| + \sum_{j=1}^{t} |D_j|}$$
$$\sim m p^{|C_i|} \mathbb{P}(\mathcal{V} \ni \mathcal{D}).$$

Because $\mathcal{C}$ is a minimal covering family, $|C_i| \geq 2$, and $m p^{|C_i|} \leq m p^2 \ll 1$, and hence $\mathbb{P}(\mathcal{V} \ni \mathcal{C} \cup \mathcal{D}) \ll \mathbb{P}(\mathcal{V} \ni \mathcal{D})$. In the latter case where $\mathcal{D}$ has a member such that $D_j \notin \mathcal{C}$, a similar reasoning shows that $\mathbb{P}(\mathcal{V} \ni \mathcal{C} \cup \mathcal{D}) \ll \mathbb{P}(\mathcal{V} \ni \mathcal{C})$. We may hence conclude that

$$\mathbb{P}(\mathcal{V} \ni \mathcal{C}, \mathcal{V} \ni \mathcal{D}) = \mathbb{P}(\mathcal{V} \ni \mathcal{C} \cup \mathcal{D}) \ll \mathbb{P}(\mathcal{V} \ni \mathcal{C}) + \mathbb{P}(\mathcal{V} \ni \mathcal{D})$$

for all distinct $\mathcal{C}, \mathcal{D} \in \text{MCF}(R)$. Therefore, the proof is completed by

$$U_2 \ll \sum_{\mathcal{C}, \mathcal{D}} \left(\mathbb{P}(\mathcal{V} \ni \mathcal{C}) + \mathbb{P}(\mathcal{V} \ni \mathcal{D}) \right) \leq 2 |\text{MCF}(R)| U_1.$$

5.2   Covering Densities of Certain Subgraphs

In order to bound the variances of subgraph counts we will use the covering densities of (partially) overlapping pairs of 2-stars and triangles. Figure 3 displays the graphs obtained as a union of two partially overlapping triangles. Figure 4 displays the graphs produced by overlapping 2-stars.

3-cycle   Diamond   Butterfly

Fig. 3. Graphs obtained as unions of overlapping triangles.

According to Theorem 3.1, the covering densities of subgraphs may be computed from their minimal covering families. For a triangle R with $V(R) = \{1, 2, 3\}$ and $E(R) = \{12, 13, 23\}$, the minimal covering families are[4] $\{123\}$ and

---

[4] For clarity, we write 12 and 123 as shorthands of the sets $\{1, 2\}$ and $\{1, 2, 3\}$.

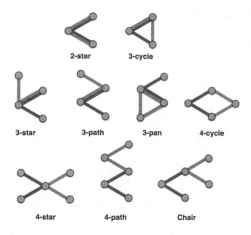

Fig. 4. Graphs obtained as unions of overlapping 2-stars.

$\{12, 13, 23\}$. The minimal covering families of the a 3-path R with $V(R) = \{1, 2, 3, 4\}$ and $E(R) = \{12, 23, 34\}$ are given by $\{1234\}$, $\{12, 234\}$, $\{123, 34\}$, and $\{12, 23, 34\}$.

The covering densities of stars are found as follows. Fix $r \geq 1$, and let R be the r-star such that $V(R) = \{1, 2, \ldots, r + 1\}$ and $E(R) = \{\{1, r + 1\}, \{2, r + 1\}, \ldots, \{r, r + 1\}\}$. The minimal covering families of R are of the form $\mathcal{C} = \{S \cup \{r + 1\} : S \in \mathcal{S}\}$, where $\mathcal{S}$ is a partition of the leaf set $\{1, \ldots, r\}$ into nonempty subsets. For any such $\mathcal{C}$ we have $|\mathcal{C}| = |\mathcal{S}|$ and $||\mathcal{C}|| = r + |\mathcal{S}|$. Hence

$$\mathbb{P}(G \supset r\text{-star}) \sim \sum_{k=1}^{r} \left\{ {r \atop k} \right\} m^k p^{k+r},$$

where $\left\{ {r \atop k} \right\}$ equals the number of partitions of $\{1, \ldots, r\}$ into k nonempty sets. These coefficients are known as Stirling numbers of the second kind [11] and can be computed via $\left\{ {r \atop k} \right\} = \frac{1}{k!} \sum_{j=0}^{k} (-1)^{k-j} \binom{k}{j} j^r$. Hence,

$$\mathbb{P}(G \supset r\text{-star}) \sim \begin{cases} mp^3 + m^2p^4, & r = 2, \\ mp^4 + 3m^2p^5 + m^3p^6, & r = 3, \\ mp^5 + 7m^2p^6 + 6m^3p^7 + m^4p^8, & r = 4. \end{cases}$$

Table 1 summarises approximate covering densities of overlapping pairs of 2-stars and triangles. The table is computed by first listing all minimal covering families of the associated subgraphs, as shown in Table 2. We also use the following observations (for $p \ll m^{-1/2} \ll 1$) to cancel some of the redundant terms in the expressions.

4-path: $m^2p^7 \ll m^2p^6$ and $m^3p^8 \ll m^3p^7$

4-cycle: $m^2p^6 \ll mp^4$

3-pan: $m^3p^7 \ll m^2p^5$

Diamond: $m^2p^6 \ll mp^4$ and $m^4p^9 \ll m^3p^7$

Butterfly: $m^5p^{12} \ll m^5p^{11}$, $m^4p^{11} \ll m^3p^8 \ll m^2p^6$, $m^3p^{10} \leq m^3p^8 \ll m^2p^6$.

Table 1. Approximate densities of some subgraphs.

R	$|V(R)|$	$|E(R)|$	Appr. density ($p \ll m^{-1/2} \ll 1$)	Appr. density ($p \sim \mu m^{-1}$)
1-star	2	1	mp^2	$\mu^2 m^{-1}$
2-star	3	2	$mp^3 + m^2 p^4$	$(1+\mu)\mu^3 m^{-2}$
3-cycle	3	3	$mp^3 + m^3 p^6$	$\mu^3 m^{-2}$
3-star	4	3	$mp^4 + 3m^2 p^5 + m^3 p^6$	$(1 + 3\mu + \mu^2)\mu^4 m^{-3}$
3-path	4	3	$mp^4 + 2m^2 p^5 + m^3 p^6$	$(1 + 2\mu + \mu^2)\mu^4 m^{-3}$
4-cycle	4	4	$mp^4 + 4m^3 p^7 + m^4 p^8$	$\mu^4 m^{-3}$
3-pan	4	4	$mp^4 + m^2 p^5 + m^4 p^8$	$(1+\mu)\mu^4 m^{-3}$
Diamond	4	5	$mp^4 + 2m^3 p^7 + m^5 p^{10}$	$\mu^4 m^{-3}$
4-star	5	4	$mp^5 + 7m^2 p^6 + 6m^3 p^7 + m^4 p^8$	$(1 + 7\mu + 6\mu^2 + \mu^3)\mu^5 m^{-4}$
4-path	5	4	$mp^5 + 3m^2 p^6 + 3m^3 p^7 + m^4 p^8$	$(1 + 3\mu + 3\mu^2 + \mu^3)\mu^5 m^{-4}$
Chair	5	4	$mp^5 + 4m^2 p^6 + 4m^3 p^7 + m^4 p^8$	$(1 + 4\mu + 4\mu^2 + \mu^3)\mu^5 m^{-4}$
Butterfly	5	6	$mp^5 + m^2 p^6 + 2m^4 p^9 + 4m^5 p^{11} + m^6 p^{12}$	$(1+\mu)\mu^5 m^{-4}$

5.3   Proofs of Theorems 3.1, 3.2, and 3.3

Proof (of Theorem 3.1). Denote $\hat{\lambda} = \hat{\lambda}(G^{(n_0)})$ and $\hat{N} = N_{K_2}(G^{(n_0)})$. Then the variance of $\hat{\lambda}$ is given by

$$\text{Var}(\hat{\lambda}) = 4\frac{n^2}{n_0^4}\text{Var}(\hat{N}). \tag{5.1}$$

By writing

$$\hat{N} = \sum_{e\in\binom{[n_0]}{2}} 1(G \supset e) \quad \text{and} \quad \hat{N}^2 = \sum_{e\in\binom{[n_0]}{2}}\sum_{e'\in\binom{[n_0]}{2}} 1(G \supset e)1(G \supset e'),$$

we find that $\mathbb{E}\hat{N} = \binom{n_0}{2}\mathbb{P}(G \supset K_2)$ and

$$\mathbb{E}\hat{N}^2 = \binom{n_0}{2}\mathbb{P}(G \supset K_2) + 2(n_0 - 2)\binom{n_0}{2}\mathbb{P}(G \supset S_2) + \binom{n_0}{2}\binom{n_0 - 2}{2}\mathbb{P}(G \supset K_2)^2.$$

Because the last term above is bounded by

$$\binom{n_0}{2}\binom{n_0 - 2}{2}\mathbb{P}(G \supset K_2)^2 \leq \binom{n_0}{2}^2\mathbb{P}(G \supset K_2)^2 = (\mathbb{E}\hat{N})^2,$$

it follows that

$$\text{Var}(\hat{N}) \leq \binom{n_0}{2}\mathbb{P}(G \supset K_2) + 2(n_0 - 2)\binom{n_0}{2}\mathbb{P}(G \supset S_2)$$

$$= (1 + o(1))\frac{1}{2}n_0^2 mp^2 + (1 + o(1))n_0^3(mp^3 + m^2 p^4).$$

Hence by (5.1),

$$\text{Var}(\hat{\lambda}) = O(n_0^{-2}n^2 mp^2) + O(n_0^{-1}n^2 mp^3) + O(n_0^{-1}n^2 m^2 p^4),$$

Table 2. Minimal covering families of the subgraphs in Figs. 3 and 4 (stars excluded).

3-path	$|\mathcal{C}|$	$||\mathcal{C}||$
1234	1	4
123, 34	2	5
234, 12	2	5
12, 23, 34	3	6

4-cycle	$|\mathcal{C}|$	$||\mathcal{C}||$
1234	1	4
123, 134	2	6
124, 234	2	6
123, 14, 34	3	7
124, 23, 34	3	7
134, 12, 23	3	7
234, 12, 14	3	7
12, 14, 23, 34	4	8

Diamond	$|\mathcal{C}|$	$||\mathcal{C}||$
1234	1	4
123, 234	2	6
123, 24, 34	3	7
234, 12, 13	3	7
124, 134, 23	3	8
124, 13, 23, 34	4	9
134, 12, 23, 24	4	9
12, 13, 23, 24, 34	5	10

3-cycle	$|\mathcal{C}|$	$||\mathcal{C}||$
123	1	3
12, 13, 23	3	6

Chair	$|\mathcal{C}|$	$||\mathcal{C}||$
12345	1	5
1234, 45	2	6
1345, 23	2	6
2345, 13	2	6
123, 345	2	6
123, 34, 45	3	7
134, 23, 45	3	7
234, 13, 45	3	7
345, 13, 23	3	7
13, 23, 34, 45	4	8

4-path	$|\mathcal{C}|$	$||\mathcal{C}||$
12345	1	5
1234, 45	2	6
2345, 12	2	6
123, 345	2	6
1245, 234	2	7
123, 34, 45	3	7
234, 12, 45	3	7
345, 12, 23	3	7
1245, 23, 34	3	8
12, 23, 34, 45	4	8

3-pan	$|\mathcal{C}|$	$||\mathcal{C}||$
1234	1	4
123, 34	2	5
134, 12, 23	3	7
234, 12, 13	3	7
12, 13, 23, 34	4	8

Butterfly	$|\mathcal{C}|$	$||\mathcal{C}||$
12345	1	5
123, 345	2	6
1234, 35, 45	3	8
1235, 34, 45	3	8
1345, 12, 23	3	8
2345, 12, 13	3	8
1245, 134, 235	3	10
1245, 135, 234	3	10
123, 34, 35, 45	4	9
345, 12, 13, 23	4	9
1245, 134, 23, 25	4	11
1245, 235, 13, 34	4	11
1245, 135, 23, 34	4	11
1245, 234, 13, 35	4	11
134, 12, 23, 35, 45	5	11
135, 12, 23, 34, 45	5	11
234, 12, 13, 35, 45	5	11
235, 12, 13, 34, 45	5	11
1245, 13, 23, 34, 35	5	12
12, 13, 23, 34, 35, 45	6	12

and by noting that $n^2 m p^2 \sim \lambda n$, $n^2 m p^3 = m^{-1/2} n^{1/2} (nmp^2)^{3/2} \sim \lambda^{3/2} m^{-1/2} n^{1/2}$ and $n^2 m p^4 = (nmp^2)^2 \sim \lambda^2$, we find that

$$\mathrm{Var}(\hat{\lambda}) = O\left(n_0^{-2} n + m^{-1/2} n_0^{-1} n^{1/2} + n_0^{-1}\right) = O\left(n_0^{-2} n + m^{-1/2} n_0^{-1} n^{1/2}\right),$$

where the last equality is true because $n_0^{-2} n \geq n_0^{-1}$. The claim now follows by Chebyshev's inequality.

Proof (of Theorems 3.2 and 3.3). The variances of N_{S_2} and N_{K_3} can be bounded from above in the same way that the variance of N_{K_2} was bounded in the proof of Theorem 3.1. The overlapping subgraphs contributing to the variance of N_{K_3} are those shown in Fig. 3. According to Table 1, the contribution of these subgraphs is $O(n_0^{|V(R)|} m^{-|V(R)|+1})$ for $|V(R)| = 3, 4, 5$, and the nonoverlapping triangles contribute $O(n_0^6 m^{-5})$. Since $\mathbb{E} N_{K_3}$ is of the order $n_0^3 m^{-2}$, it follows that $\mathrm{Var}(N_{K_3}/\mathbb{E} N_{K_3}) = o(1)$ for $n_0 \gg n^{2/3}$.

The same line of proof works for N_{S_2}, i.e., we note that the subgraphs appearing in $\mathrm{Var}(N_{S_2})$ are those shown in Fig. 4 and their contributions to the variance are listed in Table 1. Again, it follows that $\mathrm{Var}(N_{S_2}/\mathbb{E} N_{S_2}) = o(1)$ for $n_0 \gg n^{2/3}$.

Hence we may conclude using Chebyshev's inequality that

$$N_{K_3}(G^{(n_0)}) = (1 + o_p(1))\mathbb{E}N_{K_3}(G^{(n_0)}) = (1 + o_p(1))\binom{n_0}{3}\mu^3 m^{-2}$$

$$N_{S_2}(G^{(n_0)}) = (1 + o_p(1))\mathbb{E}N_{S_2}(G^{(n_0)}) = (1 + o_p(1))3\binom{n_0}{3}(1 + \mu)\mu^3 m^{-2},$$

and the claim of Theorem 3.2 follows.

Further, in the proof of Theorem 3.1 we found that

$$N_{K_2}(G^{(n_0)}) = (1 + o_p(1))\mathbb{E}N_{K_2}(G^{(n_0)}) = (1 + o_p(1))\binom{n_0}{2}\mu^2 m^{-1}.$$

Hence the claims of Theorem 3.3 follow from the above expressions combined with the continuous mapping theorem.

6   Conclusions

In this paper we discussed the estimation of parameters for a large random intersection graph model in a balanced sparse parameter regime characterised by mean degree λ and attribute intensity μ, based on a single observed instance of a subgraph induced by a set of n_0 nodes. We introduced moment estimators for λ and μ based on observed frequencies of 2-stars and triangles, and described how the estimators can be computed in time proportional to the product of the maximum degree and the number of observed nodes. We also proved that in this parameter regime the statistical network model under study has a non-trivial empirical transitivity coefficient which can be approximated by a simple parametric formula in terms of μ.

For simplicity, our analysis was restricted to binomial undirected random intersection graph models, and the statistical sampling scheme was restricted induced subgraph sampling, independent of the graph structure. Extension of the obtained results to general directed random intersection graph models with general sampling schemes is left for further study and forms a part of our ongoing work.

Acknowledgments. Part of this work has been financially supported by the Emil Aaltonen Foundation, Finland. We thank Mindaugas Bloznelis for helpful discussions, and the two anonymous reviewers for helpful comments.

References

1. Ball, F.G., Sirl, D.J., Trapman, P.: Epidemics on random intersection graphs. Ann. Appl. Probab. **24**(3), 1081–1128 (2014). http://dx.doi.org/10.1214/13-AAP942
2. Bickel, P.J., Chen, A., Levina, E.: The method of moments and degree distributions for network models. Ann. Statist. **39**(5), 2280–2301 (2011). http://dx.doi.org/10.1214/11-AOS904

3. Bloznelis, M.: The largest component in an inhomogeneous random intersection graph with clustering. Electron. J. Combin. **17**(1), R110 (2010)
4. Bloznelis, M.: Degree and clustering coefficient in sparse random intersection graphs. Ann. Appl. Probab. **23**(3), 1254–1289 (2013). http://dx.doi.org/10.1214/12-AAP874
5. Bloznelis, M., Kurauskas, V.: Clustering function: another view on clustering coefficient. J. Complex Netw. **4**, 61–86 (2015)
6. Bloznelis, M., Leskelä, L.: Diclique clustering in a directed random graph. In: Bonato, A., Graham, F.C., Pralat, P. (eds.) WAW 2016. LNCS, vol. 10088, pp. 22–33. Springer, Cham (2016). doi:10.1007/978-3-319-49787-7_3
7. Britton, T., Deijfen, M., Lagerås, A.N., Lindholm, M.: Epidemics on random graphs with tunable clustering. J. Appl. Probab. **45**(3), 743–756 (2008). http://dx.doi.org/10.1239/jap/1222441827
8. Deijfen, M., Kets, W.: Random intersection graphs with tunable degree distribution and clustering. Probab. Eng. Inform. Sc. **23**(4), 661–674 (2009). http://dx.doi.org/10.1017/S0269964809990064
9. Gleich, D.F., Owen, A.B.: Moment-based estimation of stochastic Kronecker graph parameters. Internet Math. **8**(3), 232–256 (2012). http://projecteuclid.org/euclid.im/1345581012
10. Godehardt, E., Jaworski, J.: Two models of random intersection graphs and their applications. Electron. Notes Discr. Math. **10**, 129–132 (2001)
11. Graham, R.L., Knuth, D.E., Patashnik, O.: Concrete Mathematics. Addison-Wesley, Reading (1994)
12. van der Hofstad, R.: Random Graphs and Complex Networks. Cambridge University Press, Cambridge (2017)
13. Karoński, M., Scheinerman, E.R., Singer-Cohen, K.B.: On random intersection graphs: the subgraph problem. Combin. Probab. Comput. **8**(1–2), 131–159 (1999). http://dx.doi.org/10.1017/S0963548398003459
14. Kolaczyk, E.D.: Statistical Analysis of Network Data. Springer, New York (2009)
15. Lagerås, A.N., Lindholm, M.: A note on the component structure in random intersection graphs with tunable clustering. Electron. J. Combin. **15**(1), Note 10, 8 (2008). http://www.combinatorics.org/Volume_15/Abstracts/v15i1n10.html
16. Newman, M.E.J.: The structure and function of complex networks. SIAM Rev. **45**(2), 167–256 (2003). http://dx.doi.org/10.1137/S003614450342480
17. Nikoletseas, S., Raptopoulos, C., Spirakis, P.G.: Maximum cliques in graphs with small intersection number and random intersection graphs. In: Rovan, B., Sassone, V., Widmayer, P. (eds.) MFCS 2012. LNCS, vol. 7464, pp. 728–739. Springer, Heidelberg (2012). doi:10.1007/978-3-642-32589-2_63
18. Tsourakakis, C.E.: Fast counting of triangles in large real networks without counting: algorithms and laws. In: 2008 Eighth IEEE International Conference on Data Mining, pp. 608–617, December 2008
19. Wasserman, S., Faust, K.: Social Network Analysis: Methods and Applications. Cambridge University Press, Cambridge (1994)

Common Adversaries Form Alliances: Modelling Complex Networks via Anti-transitivity

Anthony Bonato[✉], Ewa Infeld, Hari Pokhrel, and Paweł Prałat

Ryerson University, Toronto, Canada
abonato@ryerson.ca

Abstract. Anti-transitivity captures the notion that enemies of enemies are friends, and arises naturally in the study of adversaries in social networks and in the study of conflicting nation states or organizations. We present a simplified, evolutionary model for anti-transitivity influencing link formation in complex networks, and analyze the model's network dynamics. The Iterated Local Anti-Transitivity (or ILAT) model creates anti-clone nodes in each time-step, and joins anti-clones to the parent node's non-neighbor set. The graphs generated by ILAT exhibit familiar properties of complex networks such as densification, short distances (bounded by absolute constants), and bad spectral expansion. We determine the cop and domination number for graphs generated by ILAT, and finish with an analysis of their clustering coefficients. We interpret these results within the context of real-world complex networks and present open problems.

1 Introduction

Transitivity is a pervasive and folkloric notion in social networks, summarized in the adage that "friends of friends are more likely friends". A simplified, deterministic model for transitivity was posed in [3,4], where nodes are added over time, and each node's *clone* is adjacent to it and all of its neighbors. The resulting Iterated Local Transitivity (or ILT) model, while elementary to define, simulates many properties of social and other complex networks. For example, as shown in [4], graphs generated by the model densify over time, have the small world property (that is, small distances and high local clustering), and exhibit bad spectral expansion. For further properties of the ILT model, see [5,12]

Complex networks contain numerous mechanisms governing link formation, however. Structural balance theory in social network analysis cites several mechanisms to complete triads [11]. Another folkloric adage is that "enemies of enemies are more likely friends". Adversarial relationships may be modelled by non-adjacency, and so we have the resulting closure of the triad as described in Fig. 1.

Such triad closure is suggestive of an analysis of adversarial relationships between nodes as one mechanism for link formation. For instance, in social networks, we may consider both friendship ties and enmity (or rivalry) between

Research supported by grants from NSERC and Ryerson University.

A. Bonato et al. (Eds.): WAW 2017, LNCS 10519, pp. 16–26, 2017.
DOI: 10.1007/978-3-319-67810-8_2

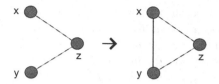

Fig. 1. Nodes x and y share z as a mutual adversary, and so form an alliance.

actors. We may also consider opposing networks of nation states or rival organizations, and consider alliances formed by mutually shared adversaries. See [10] for a recent study using the spatial location of cities to form an interaction network, where links enable the flow of cultural influence, and may be used to predict the rise of conflicts and violence. Another example comes from market graphs, where the nodes are stocks, and stocks are adjacent as a function of their correlation measured by a threshold value $\theta \in (0, 1)$. Market graphs were considered in the case of negatively correlated (or adversarial) stocks, where stocks are adjacent if $\theta < \alpha$, for some positive α; see [1].

In the present paper, we consider a simplified, deterministic model for anti-transitivity in complex networks. The Iterated Local Anti-Transitivity (or ILAT) model duplicates nodes in each time-step by forming *anti-clone* nodes, and joins them to the parent node's non-neighbor set. We give a precise definition of the model below in the next section. Perhaps unexpectedly, graphs generated by the ILAT model exhibit familiar properties of complex networks such as densification, small world properties, and bad spectral expansion (analogously to, but different from properties exhibited by ILT).

We organize the discussion in this extended abstract as follows. In Sect. 2, we give a precise definition of the ILAT model and examine its basic properties. We prove that graphs generated by ILAT densify over time. We derive the density of ILAT graphs, and consider their degree distribution. In Sect. 3, we prove that ILAT graphs have diameter 3 for sufficiently large time-steps (regardless of the initial graph). Further, we determine after several time-steps, ILAT graphs have cop number at most 2 and domination number 3. We include in Sect. 4 an analysis of the clustering coefficients and provide upper and lower bounds. The final section interprets our results within real-world complex networks, and presents open problems derived from the analysis of the model.

We consider undirected graphs throughout the paper. For background on graph theory, the reader is directed to [13]. Additional background on complex networks may be found in the book [2].

2 The ILAT Model

The Iterated Local Anti-Transitivity (or ILAT) model generates a sequence $(G_t : t \geq 0)$ of graphs over a sequence of discrete time-steps. The one parameter of the model is the initial graph G_0. Assuming the graph at time G_t is defined, we define G_{t+1} as follows. For a given node $x \in V(G_t)$, define its *anti-clone* x' as a new node adjacent to non-neighbors of x. More precisely, x' is adjacent to all nodes in $N^c(x)$, where $N^c(x) = \{y \in V(G_t) : xy \notin E(G)\}$. To form G_{t+1}, to each node x add its anti-clone x'.

The intuition behind that model is that the anti-clone x' is adversarial with x, and non-neighbors of x (that is, its own adversaries) become allied with x'. This process, therefore, iteratively applies the triad closure in Fig. 1. Note that the number of nodes doubles in each time-step, and the set of anti-clones forms an independent set. See Fig. 2 for an example.

We introduce some simplifying notation. Let n_t be the number of nodes at time t, e_t be the number of edges at time t, and the degree of a node x at time t will be denoted $\deg_t(x)$. We define the *co-degree* of x at time t as $\deg^c_t(x) = n_t - \deg_t(x) - 1$. It is straightforward to note that for $t \geq 1$, $n_t = 2n_{t-1} = 2^t n_0$. Further, for an existing node $x \in V(G_t)$,

$$\deg_{t+1}(x) = n_t - 1 \tag{1}$$
$$\deg_{t+1}(x') = \deg^c_t(x). \tag{2}$$

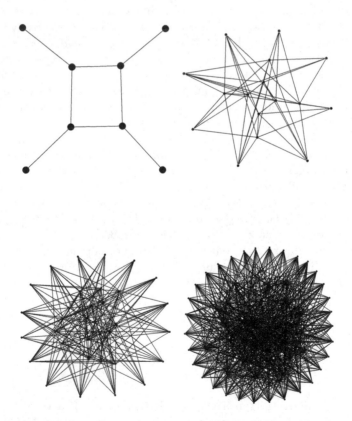

Fig. 2. An example of the first four time-steps of the ILAT model, where the initial graph is the four-cycle C_4.

The ILAT model generates graphs that densify as we prove next. While the proof is elementary, the result is not a priori obvious from the model. One interpretation is that in networks where anti-transitivity is pervasive, we expect that many alliances form in the network over time.

Theorem 1. *The ratio e_t/n_t tends to infinity with t.*

Proof. Note that by the definition of the model and (2), we have that

$$
\begin{aligned}
e_{t+1} &= e_t + \sum_{x \in V(G_t)} \deg_t{}^c(x) \\
&= e_t + n_t{}^2 - 2e_t - n_t \\
&= n_t{}^2 - e_t - n_t.
\end{aligned}
$$

Solving this recurrence, we derive that

$$
\begin{aligned}
e_t &= n_{t-1}{}^2 \left(\frac{4}{5}\right)\left(1 - \left(-\frac{1}{4}^{t-1}\right)\right) - n_{t-1}\left(\frac{2}{3}\right)\left(1 - \left(-\frac{1}{2}^{t-1}\right)\right) \\
&= 2^{2t}(n_0)^2 \left(\frac{1}{5}\right)\left(1 - \left(-\frac{1}{4}^{t-1}\right)\right)(1 - o(1)).
\end{aligned}
$$

Hence, we obtain that $e_t/n_t = \Omega(2^t)$. $\square$

Note that Theorem 1 immediately gives the limiting density of ILAT graphs. Let D_t be the density of G_t; that is, $D_t = \frac{e_t}{\binom{n_t}{2}}$.

Corollary 1. *As $t \to \infty$, we have that $D_t \to 2/5$.*

We next consider the degrees of vertices in the graph G_t. For each node x at time t, we create its anti-clone x' at time $t+1$. Then at time $t+2$ we create x'' from x and $(x')'$ from x'. For any node x that was created at a time-step $k < t$, we have directly from (1) that

$$
\deg_t(x) = \frac{n_t}{2} - 1.
$$

If $t > 1$, then of the newly created nodes, half are anti-clones x' of nodes x that have already existed at time $t-2$, and therefore, their degree at time $t-1$ was

$$
\deg_{t-1}(x) = \frac{n_{t-1}}{2} - 2 = \frac{n_t}{4} - 1.
$$

These anti-clones have at time t,

$$
\deg_t(x') = n_{t-1} - \deg_{t-1}(x) = \frac{n_t}{4} + 2.
$$

Similarly, if $t > 2$ then there are $\frac{n_t}{8}$ nodes y'' created at time t that are anti-clones of nodes y' created at time $t - 1$ from nodes y at least as old as $t - 3$. Then since by the previous argument $\deg_{t-1}(y') = \frac{n_{t-1}}{4} + 2$, we have that

$$\deg_t(y'') = \frac{3n_t}{8} - 1.$$

If we continue in this fashion, then by induction we will find that at time t, we have that $2^{-k}n_t$ nodes have degree $a_k + (-1)^{k-1}2$ provided that for $k < t$:

$$a_1 = \frac{n_t}{2} - 1,$$

and

$$a_k = \frac{1}{2} - \frac{a_{k-1}}{2}.$$

If $t > 1$, then of the newly created nodes, half are anti-clones x' of nodes x that already existed at time $t - 2$. Therefore, the degree of those nodes x at time $t - 1$ was

$$\deg_{t-1}(x) = \frac{n_{t-1}}{2} - 1 = \frac{n_t}{4} - 1.$$

Their new anti-clones x' have, at time t,

$$\deg_t(x') = n_{t-1} - \deg_{t-1}(x) = \frac{n_t}{4} + 1.$$

Similarly, if $t > 2$ then there are $\frac{n_t}{8}$ nodes y'' created at time t that are anti-clones of nodes y' created at time $t - 1$ from nodes y at least as old as $t - 3$. Then since by the previous argument $\deg_{t-1}(y') = \frac{n_{t-1}}{4} + 1$, we have that

$$\deg_t(y'') = \frac{3n_t}{8} - 1.$$

If we continue in this fashion, then by induction we will find that at time t, we have that $2^{-k}n_t$ nodes of degree $a_k + (-1)^{k-1}2$ provided that for $k < t$:

$$a_1 = \frac{n_t}{2} - 2,$$

and

$$a_k = \frac{1}{2} - \frac{a_{k-1}}{2}.$$

3   Distances and Graph Parameters

The distances within graphs generated by ILAT become very small, with diameter 3. Hence, highly anti-transitive networks exhibit short paths between nodes; this occurs at time-step $t = 2$, regardless of the starting diameter of G_0.

Theorem 2. *Let $t \geq 2$, then the diameter* $\mathrm{diam}(G_t)$ *of* G_t *is 3.*

Note that the value $t = 2$ in Theorem 2 is sharp. For example, we may take G_0 to be a path of length 4. Or we may consider an initial graph of K_3, in which case the graph at $t = 1$ is disconnected.

Proof of Theorem 2. We show first that for $t \geq 1$, the diameter of G_t is at least 3. To see this, consider the distance between some node x that existed at time $t - 1$ and its anti-clone x' created at time t. They are not adjacent and have no common neighbors, and so we have that $d(x, x') \geq 3$.

We next show that for $t \geq 2$, any two nodes that are not newly created are at most distance 2 apart. For this, let x, y be two distinct nodes that already existed at time $t - 1$. Since the node degree at time $t - 1$ is bounded by $n_t/4 - 1$, by the pigeonhole principle there is another node z that also existed at $t - 1$ that is not adjacent to either of them. Hence, z' is adjacent to both nodes and so $d(x, y) \leq 2$.

Let x', y' be two separate nodes newly anti-cloned from some nodes x, y. Since the node degree at time $t - 1$ is bounded by $\max\{0, n_t/4 - 1\}$, by the pigeonhole principle there is another node z that also existed at $t - 1$ that is not adjacent to either x or y. Then z is adjacent to both x' and y', and so $d(x', y') \leq 2$. Hence, any two nodes that both newly created are at most distance 2 apart.

The only case we have not considered are pairs of nodes where one is newly created and one is not. But if $t \geq 3$, then every newly created node has a neighbor that is not newly created and vice versa. Therefore, any such pair can be connected by a path of length at most 3.   □

The pairs of nodes we have not considered so far are ones where exactly one node is newly created, but is not a anti-clone of the other. If they are not adjacent, then we would like to know if they have a common neighbor. Let the node that already existed at time $t - 1$ be x, and the newly created node be y', cloned from some node $y \neq x$. Nodes x and y' can have a common neighbor unless the neighborhood of x at time $t - 1$ (other than possibly y itself) was a subset of the neighborhood of y at time $t - 1$ (which would be the case when $x = y$).

Theorem 3. *If x and y are nodes of G_t that are not newly created at time t, with $t \geq 2$ and $x \neq y$, and it is not the case that both x and y belonged to G_0, then $d(x, y') \leq 2$.*

Proof. Unless x and y are adjacent, we have that $d(x, y') = 1$. So suppose that x and y are adjacent. Suppose that they did not both belong to the initial graph G_0. Since they are adjacent, one of them was created later than the other. If y was created later, then every neighbor of x that was created at the same time as y is now a common neighbor of x and y'. If x was created later, but before $t - 1$, then every node adjacent to y but not x at the time produced a anti-clone of the type we need. We are left with a case where x was created at time $t - 1$, and y was created earlier.

We want to find a common neighbor of x and y' that was created at $t - 2$ or earlier. x was created at time $t - 1$, so it was cloned from a node with has either $n_t/8 - 2$, $n_t/16 + 2$ or about $n_t/12$ neighbors that already existed at time $t - 1$, and so x has either $n_t/8 + 2$, $3n_t/16 - 2$, or about $n_t/6$ neighbors older than itself. By the same argument, y' has either $n_t/8 + 2$, $3n_t/16 - 2$, or about $n_t/6$ neighbors at least as old as $t - 2$. There are in total $n_t/4$ nodes at least as old as $t - 2$. So by the pigeonhole principle, they must have such a neighbor in common.   □

Let L_t denote the average distance at time t.

Corollary 2. *The average distance L_t tends to 1.6 in t.*

Proof. Notice that the number of pairs such that both x and y belong to G_0 is negligible, so will not change the average distance limit. Of the remaining pairs of vertices, a proportion of 0.4 are adjacent and the rest are at distance 2. We can thus, conclude that

$$\lim_{t \to \infty} L_t = 1.6.$$

□

We next turn to a brief discussion of the domination and cop numbers of the ILAT graphs. As we have noticed with other parameters such as the diameter and average distance, these two parameters are bounded above by very small constants. For more on these graph parameters, see [6] (we omit their definitions here as they are well-known and owing to space constraints). As a possible interpretation of these, we note that in networks exhibiting high anti-transitivity, a few important nodes emerge (either dominating nodes, or mobile agents represented by cops) which can reach all other nodes. Such so-called *superpower* nodes organically emerge as important actors in the network.

Theorem 4. *In G_t such that $t \geq 3$, the domination number is 3.*

Proof. Let $A = \{x, x', (x')'\}$ be as follows. For any $1 \leq k \leq t-1$, let x be a node that existed at time $k - 1$ and x' be the time-k anti-clone of x. Let x'' be the time-$(k + 1)$ anti-clone of x'. Then any node of G_t not in A is either adjacent to x', adjacent to x'', or a node created at time $k + 1$ that is not adjacent to x', in which case it must be adjacent to x. Therefore, A is a dominating set of G_t.

If $t \geq 1$, then we can never find a dominating set of size 2. The node degrees are bounded by $\frac{n_t}{2} - 2$. Therefore, the union of neighborhoods of any two nodes contains at most $n_t - 4$ nodes.   □

Theorem 5. *If $t \geq 2$, then the cop number of G_t is at most 2.*

Proof. We now describe how two cops may capture the robber. Fix $v \in V(G_{t-1})$. Then each vertex of G_{t-1} is adjacent to one of v or v'. Place the cops on v and v'. Hence, the robber must begin on an anti-clone say u' newly created at time t not adjacent to either v or v'. Now there must be an x in G_t joined to u', otherwise, u is a universal vertex in G_{t-1} which is a contradiction (here is where we use $t \geq 2$). It is straightforward to show that there is a perfect matching between x, x' and v, v', and so the cops move to x, x'. The robber must move to a vertex z in G_{t-1}. But z is joined to one of x or x' and the robber is caught in the next move.   □

Note that we must have $t \geq 2$ in Theorem 5 or the cop number could be larger than 2. For example, if G_0 is a K_3, then G_1 is the disjoint union of K_3 and $\overline{K_3}$, which has cop number 4.

4   Clustering Coefficient

For a node v, define $c_t(v)$ to be the (local) clustering coefficient of the node v at time t. We note that in the ILAT model, older nodes exhibit significant local clustering over time.

Theorem 6. *Let* $k \in \mathbb{N}$. *For node* v *created at time* k, *with* $t > k$, *if* $\lim_{t \to \infty} c_t(v)$ *exists, then we have that*

$$\lim_{t \to \infty} c_t(v) = 0.4.$$

Hence, the clustering coefficient of a node v tends to 0.4 as v grows old, which matches the density of the graph.

Proof of Theorem 6. Let $c'_t(v) = c'_t$ be the density of v's non-neighbor-hood set at time t, and let $c''_t(v) = c''_t$ be the density between the neighborhood and the non-neighborhood of v. Hence, the number of edges with both endpoints in the neighborhood of v is $c_t(v)\binom{\deg_t(v)}{2}$, the number of edges with both endpoints in the non-neighborhood of v is $c'_t\binom{n_t - \deg_t(v) - 1}{2}$, the number of edges with one endpoint in the neighborhood of v, and the remaining number of edges in the non-neighborhood of v is $c''_t \deg_t(v)(n_t - \deg_t(v) - 1)$.

We write $a \sim b$ if $a = b(1 + o(1))$. For large t, we may approximate the degree by $\deg_t(v) \sim n_t - \deg_t(v) - 1 \sim \frac{n_t}{2}$. Further, since the total number of edges in the graph tends to $0.4\binom{n_t}{2}$, we have that

$$\frac{c_t + c't + 2c''_t}{4} \sim \frac{2}{5},$$

and

$$c'_t \sim \frac{8}{5} - c_t - 2c''_t.$$

Then we may determine $c_{t+1}(v) = c_{t+1}$ by counting the edges with both endpoints in the neighborhood of v at time $t + 1$. These are either the same edges that contributed to $c_t(v)$, or edges between the t-time neighborhood of v and the anti-clones of its non-neighborhood, giving the following equations:

$$c_{t+1}\binom{n_t}{2} \sim c_t\binom{n_t/2}{2} + (1 - c''_t)\frac{n_t^2}{4},$$

$$c_{t+1} \sim \frac{c_t}{4} + \frac{1 - c''_t}{2}.$$

Further, we have that

$$c''_{t+1} = \frac{c''_t}{4} + \frac{1 - c'_t}{4} + \frac{1 - c_t}{4}$$

$$c''_{t+1} = \frac{c''_t}{4} + \frac{1 - \frac{2}{5} + c_t(v) + 2c''_t}{4} + \frac{1 - c_t}{4}, \text{ and}$$

$$c''_{t+1} = \frac{3c''_t + \frac{2}{5}}{4}.$$

By hypothesis, the limiting value of c_t exists and we call this quantity c. In particular, we have that for a sufficiently large t that, $c_t(v) \sim c_{t+1} \sim c_{t+1} \sim c$. We have that

$$c_{t+2} = \frac{c_{t+1}}{4} + \frac{1 - c''_{t+1}}{2} = \frac{c_{t+1}}{4} + \frac{3}{4}\frac{1 - c''_t}{2} + \frac{1 - \frac{2}{5}}{8},$$

and so $c_{t+2} = c_{t+1} - \frac{3c_t}{16} + \frac{3}{40}$. By taking the limit as $t \to \infty$, we have that $\frac{3}{16}c = \frac{3}{40}$, and the result follows.   □

An open problem remains to prove that the limiting value of c_t exist. Further, computing the value of the clustering coefficient of G_t remains open.

5   Spectral Expansion

For a graph $G = (V, E)$ and sets of nodes $X, Y \subseteq V$, define $E(X, Y)$ to be the set of edges in G with one endpoint in X and the other in Y. For simplicity, we write $E(X) = E(X, X)$. The normalized Laplacian of a graph relates to important graph properties; see [7] for a reference. Let A denote the adjacency matrix and D denote the diagonal degree matrix of a graph G. Then the normalized Laplacian of G is $\mathcal{L} = I - D^{-1/2}AD^{-1/2}$. Let $0 = \lambda_0 \leq \lambda_1 \leq \cdots \leq \lambda_{n-1} \leq 2$ denote the eigenvalues of $\mathcal{L}$. The *spectral gap* of the normalized Laplacian is defined as

$$\lambda = \max\{|\lambda_1 - 1|, |\lambda_{n-1} - 1|\}.$$

A spectral gap bounded away from zero is an indication of bad expansion properties, which is characteristic for social networks; see [9]. The next theorem represents a drastic departure from the good expansion found in binomial random graphs, where $\lambda = o(1)$; see [7,8].

Theorem 7. *If λ_t is the spectral gap of G_t, then $\lambda_t \geq 3/5 + o(1)$.*

To prove Theorem 7, we use the expander mixing lemma for the normalized Laplacian (see [7] for its proof). For sets of nodes X and Y we use the notation $\text{vol}(X) = \sum_{v \in X} \deg(v)$ for the volume of X, $\bar{X} = V \setminus X$ for the complement of X, and, $e(X, Y)$ for the number of edges with one end in each of X and Y. (Note that $X \cap Y$ does not have to be empty; in general, $e(X, Y)$ is defined to be the number of edges between $X \setminus Y$ to Y plus twice the number of edges that contain only nodes of $X \cap Y$. In particular, $e(X, X) = 2|E(X)|$.)

Lemma 1. *For all sets* $X \subseteq V(G_t)$,

$$\left| e(X, X) - \frac{(\mathrm{vol}(X))^2}{\mathrm{vol}(G_t)} \right| \leq \lambda_t \frac{\mathrm{vol}(X)\mathrm{vol}(\bar{X})}{\mathrm{vol}(G_t)}.$$

Proof of Theorem 7. Let X be the set of $n_t/2$ the youngest nodes. Since X induces an independent set, we note that $e(X, X) = 0$. We derive that

$$\mathrm{vol}(G_t) \sim 2n_t{}^2/5,$$
$$\mathrm{vol}(\bar{X}) \sim n_t{}^2/4, \quad \text{and}$$
$$\mathrm{vol}(X) = \mathrm{vol}(G_t) - \mathrm{vol}(\bar{X}) \sim 3n_t{}^2/20,$$

where the second expression holds as $(n_t/2)$-many of the oldest nodes have degree $\sim n_t/2$. Hence, by Lemma 1, we have that

$$\lambda_t \geq \frac{(\mathrm{vol}(X))^2}{\mathrm{vol}(G_t)} \cdot \frac{\mathrm{vol}(G_t)}{\mathrm{vol}(X)\mathrm{vol}(\bar{X})} = \frac{\mathrm{vol}(X)}{\mathrm{vol}(\bar{X})} \sim 3/5,$$

and the proof follows. □

6   Discussion and Future Work

We introduced the Iterated Local Anti-Transitivity (ILAT) model for complex networks and analyzed properties of the graphs it generates. We proved that graphs generated by ILAT densify over time, have diameter 3, and have density tending to 0.4. ILAT graphs have small dominating sets and low cop number. We analyzed the clustering coefficient of ILAT graphs, and noted that while older nodes show high (local) clustering, the (global) clustering coefficient is less than what is expected in binomial random graphs with the same expected degree. In addition, we showed that graphs generated by ILAT exhibit bad spectral expansion as found in social networks.

Theoretical results presented here for the ILAT model are suggestive of several emergent properties in networks where anti-transitivity governs link formation. For instance, the presence of small (3-element) dominating sets suggest the emergence of nodes we describe as *superpowers*, which have broad influence in the network. Such nodes may emerge naturally in real-world networks which are highly anti-transitive, owing to a high number of alliances against common adversaries. Similarly, the presence of short paths, high density, and high (local) clustering of older nodes in ILAT graphs suggests that networks, where common adversaries forge alliances, naturally form tight-knit communities that are well-connected. In the sequel, it would be interesting to empirically test these hypotheses with real-world networked data.

Besides applications of the ILAT model, it raises a number of interesting graph-theoretic questions. An open problem remains to compute the clustering coefficient for ILAT graphs. Another question is to determine the induced subgraph structure of such graphs. A characterization of the induced subgraphs of ILAT graphs (that is, to determine its *age*) remains open. For example, do all finite trees appear as induced subgraphs of ILAT graphs?

References

1. Boginski, V., Butenko, S., Pardalos, P.M.: On structural properties of the market graph. In: Nagurney, A. (ed.) Innovation in Financial and Economic Networks, pp. 29–45. Edward Elgar Publishers (2003)
2. Bonato, A.: A Course on the Web Graph. Graduate Studies Series in Mathematics. American Mathematical Society, Providence (2008)
3. Bonato, A., Hadi, N., Prałat, P., Wang, C.: Dynamic models of on-line social networks. In: Proceedings of WAW 2009 (2009)
4. Bonato, A., Hadi, N., Horn, P., Prałat, P., Wang, C.: Models of on-line social networks. Internet Math. **6**, 285–313 (2011)
5. Bonato, A., Janssen, J., Roshanbin, E.: How to burn a graph. Internet Math. **1–2**, 85–100 (2016)
6. Bonato, A., Nowakowski, R.J.: The Game of Cops and Robbers on Graphs. American Mathematical Society, Providence (2011)
7. Chung, F.R.K.: Spectral Graph Theory. American Mathematical Society, Providence (1997)
8. Chung, F.R.K., Lu, L.: Complex Graphs and Networks. American Mathematical Society, U.S.A. (2004)
9. Estrada, E.: Spectral scaling and good expansion properties in complex networks. Europhys. Lett. **73**, 649–655 (2006)
10. Guo, W., Lu, X., Donate, G.M., Johnson, S.: The spatial ecology of war and peace, Preprint (2017)
11. Sack, G.: Character networks for narrative generation. In: Intelligent Narrative Technologies: Papers from the 2012 AIIDE Workshop, AAAI Technical Report WS-12-14 (2012)
12. Small, L., Mason, O.: Information diffusion on the iterated local transitivity model of online social networks. Discrete Appl. Math. **161**, 1338–1344 (2013)
13. West, D.B.: Introduction to Graph Theory, 2nd edn. Prentice Hall, Upper Saddle River (2001)

Kernels on Graphs as Proximity Measures

Konstantin Avrachenkov[1]([⊠]), Pavel Chebotarev[2], and Dmytro Rubanov[1]

[1] Inria Sophia Antipolis, Valbonne, France
{k.avrachenkov,dmytro.rubanov}@inria.fr
[2] RAS Institute of Control Sciences, Moscow, Russia
pavel4e@gmail.com

Abstract. Kernels and, broadly speaking, similarity measures on graphs are extensively used in graph-based unsupervised and semi-supervised learning algorithms as well as in the link prediction problem. We analytically study proximity and distance properties of various kernels and similarity measures on graphs. This can potentially be useful for recommending the adoption of one or another similarity measure in a machine learning method. Also, we numerically compare various similarity measures in the context of spectral clustering and observe that normalized heat-type similarity measures with log modification generally perform the best.

1 Introduction

Many graph-based semi-supervised learning methods, see e.g., [1–3,7,23,24,43] and references therein, can be viewed as the methods comparing some *distances* or *similarity measures* from unlabelled nodes to the labelled ones. An unlabelled node is attributed to a class whose labelled nodes are closer with respect to distances or similarity measures. Also, unsupervised machine learning methods such as K-means and its numerous variations are based on grouping points in a metric space, see e.g. [19,32,44]. While the plain K-means method discovers only linear boundaries between clusters in a metric space, kernel K-means methods have better sensitivity and can discover clusters of more general shapes. In addition, some kernel K-means methods are equivalent to spectral clustering [19]. A choice of a kernel may have significant impact on the clustering quality. Moreover, the distance property of the kernels can be exploited for quick grouping of points in the K-means methods [18,19]. Similarity measures are also used in the link prediction problem [5,33].

Most but not all similarity measures are defined with the help of kernels on graphs, i.e., positive semidefinite matrices with indices corresponding to the nodes. Note that according to Schoenberg's theorem [34,35] one can always transform a positive semidefinite matrix to a set of points in an Euclidian space. In contrast, the proximity property [13] is much more subtle and not all kernels on graphs appear to be proximity measures.

In this paper, we analyse distance and proximity properties of the similarity measures and kernels on graphs. All similarity measures and kernels that we

© Springer International Publishing AG 2017
A. Bonato et al. (Eds.): WAW 2017, LNCS 10519, pp. 27–41, 2017.
DOI: 10.1007/978-3-319-67810-8_3

study are defined in terms of one of the following three basic matrices: weighted adjacency matrix, combinatorial Laplacian and (stochastic) Markov matrix. We compare similarity measures and kernels on graphs both theoretically and by numerical experiments in the context of spectral clustering on the stochastic block model. We hope that our analysis will be useful for recommending the adoption of one or another similarity measure in a machine learning method. It was interesting to observe that in the context of the spectral clustering, the normalized heat-type kernels with logarithmic transformation perform the best on the stochastic block model.

2   Definitions and Preliminaries

The *weighted adjacency matrix* $W = (w_{ij})$ of a weighted undirected graph G with vertex set $V(G) = \{1, \ldots, n\}$ is the matrix with elements

$$w_{ij} = \begin{cases} \text{weight of edge } (i, j), \text{ if } i \sim j, \\ 0, \qquad\qquad\qquad\qquad \text{otherwise.} \end{cases}$$

In what follows, G is connected.

The ordinary (or combinatorial) *Laplacian matrix* L of G is defined as follows: $L = D - W$, where $D = \mathrm{Diag}(W \cdot \mathbf{1})$ is the degree matrix of G, $\mathrm{Diag}(\boldsymbol{x})$ is the diagonal matrix with vector $\boldsymbol{x}$ on the main diagonal, and $\mathbf{1} = (1, \ldots, 1)^T$. In most cases, the dimension of $\mathbf{1}$ is clear from the context.

Informally, given a weighted graph G, a *similarity measure* on the set of its vertices $V(G)$ is a function $\kappa \colon V(G) \times V(G) \to \mathbb{R}$ that characterizes similarity (or affinity, or closeness) between the vertices of G in a meaningful manner and thus is intuitively and practically adequate for empirical applications [2,18,24,33].

A *kernel on graph* is graph similarity measure that has an inner product representation. Inner product matrices (also called Gram matrices) with real entries are symmetric positive semidefinite matrices. On the other hand, any semidefinite matrix has a representation as a Gram matrix with respect to the Euclidean inner product [25].

We note that following [31,39] we prefer to write *kernel on graph* rather than *graph kernel*, as the notion of "graph kernel" refers to a kernel between graphs [41].

A *proximity measure* (or simply *proximity*) [13] on a finite set A is a function $\kappa \colon A \times A \to \mathbb{R}$ that satisfies the *triangle inequality for proximities*, viz.: for any $x, y, z \in A$, $\kappa(x, y) + \kappa(x, z) - \kappa(y, z) \le \kappa(x, x)$, and if $z = y$ and $y \ne x$, then the inequality is strict.

A proximity κ is a Σ-*proximity* ($\Sigma \in \mathbb{R}$) if it satisfies the *normalization condition*: $\sum_{y \in A} \kappa(x, y) = \Sigma$ for any $x \in A$.

By setting $z = x$ in the triangle inequality for proximities and using the arbitrariness of x and y one verifies that any proximity satisfies *symmetry*: $\kappa(x, y) = \kappa(y, x)$ for any $x, y \in A$.

Furthermore, any Σ-proximity has the *egocentrism* property: $\kappa(x, x) > \kappa(x, y)$ for any distinct $x, y \in A$ [13]. If $\kappa(x, y)$ is represented by a matrix

$K = (K_{xy}) = (\kappa(x,y))$, then egocentrism of $\kappa(x,y)$ amounts to the *entrywise diagonal dominance* of K.

If $\boldsymbol{x}_i$ and $\boldsymbol{x}_j$ are two points in the Euclidean space $\mathbb{R}^n$, then $||\boldsymbol{x}_i - \boldsymbol{x}_j||_2^2$ is the squared distance between $\boldsymbol{x}_i$ and $\boldsymbol{x}_j$. Schoenberg's theorem establishes a connection between positive semidefinite matrices (kernels) and matrices of Euclidean distances.

Theorem 1. ([34,35]). *Let K be an $n \times n$ symmetric matrix. Define the matrix*

$$\mathcal{D} = (d_{ij}) = \frac{1}{2}\big(\operatorname{diag}(K) \cdot \boldsymbol{1}^T + \boldsymbol{1} \cdot \operatorname{diag}(K)^T\big) - K, \tag{1}$$

where $\operatorname{diag}(K)$ is the vector consisting of the diagonal entries of K. Then there exists a set of vectors $\boldsymbol{x}_1, \ldots, \boldsymbol{x}_n \in \mathbb{R}^n$ such that $d_{ij} = ||\boldsymbol{x}_i - \boldsymbol{x}_j||_2^2$ $(i,j = 1, \ldots, n)$ if and only if K is positive semidefinite.

In the case described in Theorem 1, K is the Gram matrix of $\boldsymbol{x}_1, \ldots, \boldsymbol{x}_n$. Given K, these vectors can be obtained as the columns of the unique positive semidefinite real matrix B such that $B^2 = B^T B = K$. B has the expression $B = U\Lambda^{1/2}U^*$, where $\Lambda = \operatorname{Diag}(\lambda_1, \ldots, \lambda_n)$, $\Lambda^{1/2} = \operatorname{Diag}(\lambda_1^{1/2}, \ldots, \lambda_n^{1/2})$, and $A = U\Lambda U^*$ is the unitary decomposition of A [25, Corollary 7.2.11].

Connections between proximities and distances are established in [13].

Theorem 2. *For any proximity κ on a finite set A, the function*

$$d(x,y) = \frac{1}{2}(\kappa(x,x) + \kappa(y,y)) - \kappa(x,y), \quad x,y \in A \tag{2}$$

is a distance function $A \times A \to \mathbb{R}$.

This theorem follows from the proof of Proposition 3 in [13].

Corollary 1. *Let $\mathcal{D} = (d_{xy})$ be obtained by (1) from a square matrix K. If $\mathcal{D}$ has negative entries or $\sqrt{d_{xy}} + \sqrt{d_{yz}} < \sqrt{d_{xz}}$ for some $x,y,z \in \{1, \ldots, n\}$, then the function $\kappa(x,y) = K_{xy}$, $x,y \in \{1, \ldots, n\}$ is not a proximity.*

Proof. If $\sqrt{d_{xy}} + \sqrt{d_{yz}} < \sqrt{d_{xz}}$, then $d_{xy} + d_{yz} + 2\sqrt{d_{xy}d_{yz}} < d_{xz}$, i.e., the function $d(x,y) = d_{xy}$ violates the ordinary triangle inequality. Thus, it is not a distance, as well as in the case where $\mathcal{D}$ has negative entries. Hence, by Theorem 2, κ is not a proximity. □

The following theorem describes a one-to-one correspondence between distances and Σ-proximities with a fixed Σ on the same finite set.

Theorem 3. ([13]). *Let $\boldsymbol{S}$ and $\boldsymbol{D}$ be the set of Σ-proximities on A ($|A| = n$; $\Sigma \in \mathbb{R}$ is fixed) and the set of distances on A, respectively. Consider the mapping $\psi(\kappa)$ defined by (2) and the mapping $\varphi(d)$ defined by*

$$\kappa(x,y) = d(x,\cdot) + d(y,\cdot) - d(x,y) - d(\cdot,\cdot) + \frac{\Sigma}{n}, \tag{3}$$

where $d(x,\cdot) = \frac{1}{n}\sum_{y \in A} d(x,y)$ and $d(\cdot,\cdot) = \frac{1}{n^2}\sum_{y,z \in A} d(y,z)$. Then $\psi(\boldsymbol{S}) = \boldsymbol{D}$, $\varphi(\boldsymbol{D}) = \boldsymbol{S}$, and $\varphi(\psi(\kappa)), \kappa \in \boldsymbol{S}$ and $\psi(\varphi(d)), d \in \boldsymbol{D}$ are identity transformations.

Remark 1. The $K \to \mathcal{D}$ transformation (1) is the matrix form of (2). The matrix form of (3) is

$$K = -H\mathcal{D}H + \Sigma J, \tag{4}$$

where $J = \frac{1}{n}\mathbf{1}\cdot\mathbf{1}^T$ and $H = I - J$ is the *centering matrix*.

3   Kernel, Proximity and Distance Properties

3.1   Adjacency Matrix Based Kernels and Measures

Let us consider several kernels on graphs based on the weighted adjacency matrix W of a graph.

Katz Kernel. The *Katz kernel* [28] (also referred to as walk proximity [14] and von Neumann[1] diffusion kernel [37,38]) is defined[2] as follows:

$$K^{\text{Katz}}(\alpha) = \sum_{k=0}^{\infty}(\alpha W)^k = [I - \alpha W]^{-1},$$

with $0 < \alpha < (\rho(W))^{-1}$, where $\rho(W)$ is the spectral radius of W.

It is easy to see that $[I - \alpha W]$ is an M-matrix[3], i.e., a matrix of the form $A = qI - B$, where $B = (b_{ij})$ with $b_{ij} \geq 0$ for all $1 \leq i,j \leq n$, while q exceeds the maximum of the moduli of the eigenvalues of B (in the present case, $q = 1$). Thus, $[I - \alpha W]$ is a symmetric M-matrix, i.e., a Stieltjes matrix. Consequently, $[I - \alpha W]$ is positive definite and so is $K^{\text{Katz}}(\alpha) = [I - \alpha W]^{-1}$. Thus, by Schoenberg's theorem, K^{Katz} can be transformed by (1) into a matrix of squared Euclidean distances.

Moreover, the Katz kernel has the following properties:
If $[I - \alpha W]$ is row diagonally dominant, i.e., $|1 - \alpha w_{ii}| \geq \alpha \sum_{j \neq i}|w_{ij}|$ for all $i \in V(G)$ (by the finiteness of the underlying space, one can always choose α small enough such that this inequality becomes valid) then

- $K^{\text{Katz}}(\alpha)$ satisfies the triangle inequality for proximities (see Corollary 6.2.5 in [29]), therefore, transformation (2) provides a distance on $V(G)$;
- $K^{\text{Katz}}(\alpha)$ satisfies egocentrism (i.e., *entrywise* diagonal dominance; see also Metzler's property in [29]).

Thus, in the case of row diagonal dominance of $[I - \alpha W]$, the Katz kernel is a non-normalized proximity.

---

[1] M. Saerens [36] has remarked that a more suitable name could be *Neumann diffusion kernel*, referring to the *Neumann series* $\sum_{k=0}^{\infty}T^k$ (where T is an operator) named after Carl Gottfried Neumann, while a connection of that to John von Neumann is not obvious (the concept of von Neumann kernel in group theory is essentially different).

[2] In fact, L. Katz considered $\sum_{k=1}^{\infty}(\alpha W)^k$.

[3] For the properties of M-matrices, we refer to [29].

Communicability Kernel. The *communicability kernel* [20, 21, 23] is defined as follows:

$$K^{\mathrm{comm}}(t) = \exp(tW) = \sum_{k=0}^{\infty} \frac{t^k}{k!} W^k.$$

(We shall use letter "t" whenever some notion of time can be attached to the kernel parameter; otherwise, we shall keep using letter "α".) It is an instance of symmetric exponential diffusion kernels [31]. Since K^{comm} is positive semi-definite, by Schoenberg's theorem, it can be transformed by (1) into a matrix of squared Euclidean distances. However, this does not imply that K^{comm} is a proximity.

In fact, it is easy to verify that for the graph G with adjacency matrix

$$W = \begin{pmatrix} 0 & 2 & 0 & 0 \\ 2 & 0 & 1 & 0 \\ 0 & 1 & 0 & 2 \\ 0 & 0 & 2 & 0 \end{pmatrix}, \tag{5}$$

$K^{\mathrm{comm}}(1)$ violates the triangle inequality for proximities on the triple of vertices $(1, \mathbf{2}, 3)$ (the "x" element of the inequality is given in bold). On the other hand, $K^{\mathrm{comm}}(t) \to I$ as $t \to 0$, which implies that $K^{\mathrm{comm}}(t)$ with a sufficiently small t is a [non-normalized] proximity.

Note that the graph corresponding to (5) is a weighted path 1–2–3–4, and immediate intuition suggests the inequality $d(1,2) < d(1,3) < d(1,4)$ for a distance on its vertices. However, $K^{\mathrm{comm}}(3)$ induces a Euclidean distance for which $d(1,3) > d(1,4)$. For $K^{\mathrm{comm}}(4.5)$ we even have $d(1,2) > d(1,4)$. However, $K^{\mathrm{comm}}(t)$ with a small enough positive t satisfies the common intuition.

By the way, the Katz kernel behaves similarly: when $\alpha > 0$ is sufficiently small, it holds that $d(1,2) < d(1,3) < d(1,4)$, but for $\alpha > 0.375$, we have $d(1,3) > d(1,4)$. Moreover, if $0.38795 < \alpha < (\rho(W))^{-1}$, then $d(1,2) > d(1,4)$ is true.

Double-Factorial Similarity. The *double-factorial similarity* [22] is defined as follows:

$$K^{\mathrm{df}}(t) = \sum_{k=0}^{\infty} \frac{t^k}{k!!} W^k.$$

As distinct from the communicability measure, K^{df} is not generally a kernel. Say, for the graph with weighted adjacency matrix (5), $K^{\mathrm{df}}(1)$ has two negative eigenvalues. Therefore K^{df} does not generally induce a set of points in $\mathbb{R}^n$, nor does it induce a natural Euclidean distance on $V(G)$.

Furthermore, in this example, matrix $\mathcal{D}$ obtained from $K^{\mathrm{df}}(1)$ by (1) has negative entries. Therefore, by Corollary 1, the function $\kappa(x,y) = K^{\mathrm{df}}_{xy}(1)$, $x, y \in V(G)$ is not a proximity.

However, as well as $K^{\mathrm{comm}}(t)$, $K^{\mathrm{df}}(t) \to I$ as $t \to 0$. Consequently, all eigenvalues of $K^{\mathrm{df}}(t)$ converge to 1, and hence, $K^{\mathrm{df}}(t)$ with a sufficiently small positive t satisfies the triangle inequality for proximities. Thus, $K^{df}(t)$ with a small enough positive t is a kernel and a [non-normalized] proximity.

3.2 Laplacian Based Kernels and Measures

Heat Kernel. The *heat kernel* is a symmetric exponential diffusion kernel [31] defined as follows:

$$K^{heat}(t) = \exp(-tL) = \sum_{k=0}^{\infty} \frac{(-t)^k}{k!} L^k,$$

where L is the ordinary Laplacian matrix of G.

$K^{heat}(t)$ is positive-definite for all values of t, and hence, it is a kernel. Then, by Schoenberg's theorem, K^{heat} induces a Euclidean distance on $V(G)$. For our example (5), this distance for all $t > 0$ obeys the intuitive inequality $d(1,2) < d(1,3) < d(1,4)$.

On the other hand, K^{heat} is not generally a proximity. E.g., for the example (5), $K^{heat}(t)$ violates the triangle inequality for proximities on the triple of vertices $(1, \mathbf{2}, 3)$ whenever $t > 0.431$. As well as for the communicability kernel, $K^{heat}(t)$ with a small enough t is a proximity. Moreover, it is an 1-proximity, as L has row sums 0, while $L^0 = I$ has row sums 1. Thus, the 1-normalization condition is satisfied for any $t > 0$.

Normalized Heat Kernel. The *normalized heat kernel* is defined as follows:

$$K^{n\text{-}heat}(t) = \exp(-t\mathcal{L}) = \sum_{k=0}^{\infty} \frac{(-t)^k}{k!} \mathcal{L}^k,$$

where $\mathcal{L} = D^{-1/2}LD^{-1/2}$ is the normalized Laplacian, D being the degree matrix of G [15].

For this kernel, the main conclusions are the same as for the standard heat kernel. For the example (5), $K^{heat}(t)$ violates the triangle inequality for proximities on the triple of vertices $(1, \mathbf{2}, 3)$ when $t > 1.497$. It is curious to observe that the triangle inequality of the example (5) is violated starting with a larger value of t in comparison with the case of the standard heat kernel. An important distinction is that generally, $\mathcal{L}$ has nonzero row sums. As a result, $K^{n\text{-}heat}$ does not satisfy the normalization condition, and even for small t, $K^{n\text{-}heat}$ is a non-normalized proximity.

Regularized Laplacian Kernel. The *regularized Laplacian kernel*, or *forest kernel* is defined [11] as follows:

$$K^{regL}(t) = [I + tL]^{-1},$$

where $t > 0$.

As was shown in [12,14], the regularized Laplacian kernel is a 1-proximity and a row stochastic matrix. Since $[I + tL]$ is positive definite, so is $[I + tL]^{-1}$, and by Schoenberg's theorem, K^{regL} induces a Euclidean distance on $V(G)$.

For the example (5), the induced distances corresponding to K^{regL} always satisfy $d(1,2) < d(1,3) < d(1,4)$. Regarding the other properties of K^{regL}, we refer to [3,12].

It is the first encountered example of similarity measure that satisfies the both distance and proximity properties for all values of the kernel parameter.

Absorption Kernel. The *absorption kernel* [27] is defined as follows:

$$K^{\mathrm{absorp}}(t) = [tA + L]^{-1}, \quad t > 0,$$

where $A = \mathrm{Diag}(\boldsymbol{a})$ and $\boldsymbol{a} = (a_1, \ldots, a_n)^T$ is called the *vector of absorption rates* and has positive components. As $K^{\mathrm{absorp}}(t^{-1}) = t(A + tL)^{-1}$, this kernel is actually a generalization of the previous one.

Since $[tA + L]$ is positive definite, Schoenberg's theorem attaches a matrix of squared Euclidean distances to $K^{\mathrm{absorp}}(t)$.

$[tA + L]$ is a row diagonally dominant Stieltjes matrix, hence, by Corollary 6.2.5 in [29] we conclude that K^{absorp} satisfies the triangle inequality for proximities, i.e., K^{absorp} is a proximity (but not generally a Σ-proximity).

3.3   Markov Matrix Based Kernels and Measures

Personalized PageRank. *Personalized PageRank* (PPR) similarity measure is defined as follows:

$$K^{\mathrm{PPR}}(\alpha) = [I - \alpha P]^{-1},$$

where $P = D^{-1}W$ is a row stochastic (Markov) matrix, D is the degree matrix of G, and $0 < \alpha < 1$, which corresponds to the standard random walk on the graph.

In general, $K^{\mathrm{PPR}}(\alpha)$ is not symmetric, so it is not positive semidefinite, nor is it a proximity.

Moreover, the functions $d(x, y)$ obtained from K^{PPR} by transformation[4]

$$d(x, y) = \frac{1}{2}(\kappa(x, x) + \kappa(y, y) - \kappa(x, y) - \kappa(y, x)) \qquad \text{*} \qquad (6)$$

need not generally be distances. Say, for

$$W = \begin{pmatrix} 0 & 2 & 0 & 0 & 0 \\ 2 & 0 & 1 & 0 & 0 \\ 0 & 1 & 0 & 1 & 0 \\ 0 & 0 & 1 & 0 & 2 \\ 0 & 0 & 0 & 2 & 0 \end{pmatrix} \qquad (7)$$

---

[4] If K is symmetric, then (6) coincides with (2).

with $K^{\text{PPR}}(\alpha)$, one has $d(1,3) + d(3,4) < d(1,4)$ whenever $\alpha > 0.9515$.

K^{PPR} has only positive eigenvalues. However, its symmetrized counterpart $\frac{1}{2}(K^{\text{PPR}} + (K^{\text{PPR}})^T)$ may have a negative eigenvalue (say, with $\alpha \geq 0.984$ for (5) or with $\alpha \geq 0.98$ for (7)). Thus, it need not be positive semidefinite and, consequently, by Theorem 1, $\mathcal{D}$ obtained from it by (1) (or from K^{PPR} by (6)) is not generally a matrix of squared Euclidean distances.

K^{PPR} satisfies the normalization condition. For a small enough α, it can be transformed (as well as K^{comm} and K^{df}) into a distance matrix using (6).

On the other hand, one can slightly modify Personalized PageRank so it becomes a proximity. Rewrite K^{PPR} as follows:

$$[I - \alpha D^{-1}W]^{-1} = [D - \alpha W]^{-1}D.$$

Then, consider

Modified Personalized PageRank.

$$K^{\text{modifPPR}}(\alpha) = [I - \alpha D^{-1}W]^{-1}D^{-1} = [D - \alpha W]^{-1}, \quad 0 < \alpha < 1,$$

which becomes a non-normalized proximity by Corollary 6.2.5 in [29]. In particular, the triangle inequality becomes

$$\frac{K_{ii}^{\text{PPR}}(\alpha)}{d_i} - \frac{K_{ji}^{\text{PPR}}(\alpha)}{d_i} - \frac{K_{ik}^{\text{PPR}}(\alpha)}{d_k} + \frac{K_{jk}^{\text{PPR}}(\alpha)}{d_k} \geq 0,$$

which looks like an interesting inequality for Personalized PageRank. Due to symmetry, $K_{ij}^{\text{modifPPR}} = K_{ji}^{\text{modifPPR}}$, and we obtain an independent proof of the following identity for Personalized PageRank [2]:

$$\frac{K_{ij}^{\text{PPR}}(\alpha)}{d_j} = \frac{K_{ji}^{\text{PPR}}(\alpha)}{d_i}.$$

Note that replacing the Laplacian matrix $L = D - W$ with $D - \alpha W$ is a kind of alternative regularization of L. Being diagonally dominant,

$$D - \alpha W = \bar{d}I - (\bar{d}I - D + \alpha W) \tag{8}$$

(where $\bar{d}$ is the maximum degree of the vertices of G) is a Stieltjes matrix. Consequently, $D - \alpha W$ is positive definite and so is $K^{\text{modifPPR}}(\alpha) = [D - \alpha W]^{-1}$. Thus, by Schoenberg's theorem, K^{modifPPR} can be transformed by (1) into a matrix of squared Euclidean distances.

We note that Personalized PageRank can be generalized by using non-homogeneous restart [4], which will lead to the discrete-time analog of the absorption kernel. However, curiously enough, the discrete-time version has a smaller number of proximity-distance properties than the continuous-time version.

PageRank Heat Similarity Measure. *PageRank heat similarity measure* [16] is defined as follows:

$$K^{\text{heatPPR}}(t) = \exp(-t(I - P)).$$

Basically, the properties of this measure are similar to those of the standard Personalized PageRank. Say, for the example (7) with K^{heatPPR}, one has $d(1,2) + d(2,3) < d(1,3)$ whenever $t > 1.45$.

3.4 Logarithmic Similarity Measures and Transitional Properties

Given a strictly positive similarity measure $s(x, y)$, the function $\kappa(x, y) = \ln s(x, y)$ is the corresponding *logarithmic similarity*.

Using Theorem 2 it can be verified [8] that whenever $S = (s_{ij}) = (s(i,j))$ produces a strictly positive *transitional measure* on G (i.e., $s_{ij} s_{jk} \le s_{ik} s_{jj}$ for all vertices i, j, and k, while $s_{ij} s_{jk} = s_{ik} s_{jj}$ if and only if every path from i to k visits j), we haves that the logarithmic similarity $\kappa(x, y) = \ln s(x, y)$ produces a *cutpoint additive distance*, viz., a distance that satisfies $d(i,j) + d(j,k) = d(i,k)$ iff every path from i to k visits j:

$$d(i,j) = \tfrac{1}{2}(\kappa(i,i) + \kappa(j,j) - \kappa(i,j) - \kappa(j,i)) = \ln \sqrt{\frac{s(i,i)s(j,j)}{s(i,j)s(j,i)}}. \qquad (9)$$

In the case of digraphs, five transitional measures were indicated in [8], namely, *connection reliability*, *path accessibility* with a sufficiently small parameter, *walk accessibility*, and two versions of *forest accessibility*; the undirected counterparts of the two latter measures were studied in [10] and [9], respectively.

Proposition 1. K^{absorp}, K^{PPR}, *and* K^{modifPPR} *produce transitional measures.*

Proof. For $K^{\text{absorp}}(t) = [tA + L]^{-1}$, let $h = \max_i \{a_i t + d_i - w_{ii}\}$, where d_i is the degree of vertex i. Then $K^{\text{absorp}}(t) = [hI - (hI - tA - D + W)]^{-1} = [I - W']^{-1} h^{-1}$, where $W' = h^{-1}(hI - tA - D + W)$ is nonnegative with row sums less than 1. Hence, $K^{\text{absorp}}(t)$ is positively proportional to the matrix $[I - W']^{-1}$ of walk weights of the graph with weighted adjacency matrix W'.

Similarly, by (8), $K^{\text{modifPPR}}(\alpha) = [D - \alpha W]^{-1} = [I - W'']^{-1} \bar{d}^{-1}$, where $W'' = \bar{d}^{-1}(\bar{d}I - D + \alpha W)$ is nonnegative with row sums less than 1. Consequently, $K^{\text{modifPPR}}(\alpha)$ is proportional to the matrix of walk weights of the graph whose weighted adjacency matrix is W''.

Finally, $K^{\text{PPR}}(\alpha)$ is the matrix of walk weights of the digraph with weighted adjacency matrix αP.

Since by [8, Theorem 6], any finite matrix of walk weights of a weighted digraph produces a transitional measure, so do K^{absorp}, K^{PPR}, and K^{modifPPR}. $\square$

Thus, as by Proposition 1 and the results of [8], K^{Katz}, K^{regL}, K^{absorp}, K^{PPR}, and K^{modifPPR} produce transitional measures, we have that the corresponding *logarithmic* dissimilarities (9) are cutpoint additive distances.

Furthermore, if $S = (s_{ij}) = (s(i, j))$ produces a strictly positive transitional measure on G, then, obviously, $\kappa(x, y) = \ln s(x, y)$ satisfies $\kappa(y, x) + \kappa(x, z) - \kappa(y, z) \leq \kappa(x, x)$, which coincides[5] with the triangle inequality for proximities whenever $s(x, y)$ is symmetric. Therefore, as K^{Katz}, K^{regL}, K^{absorp}, and K^{modifPPR} are symmetric, we obtain that the corresponding logarithmic similarities $\kappa(x, y) = \ln s(x, y)$ are proximities.

K^{PPR} is not generally symmetric, however, it can be observed that $\widetilde{K}^{\text{PPR}}$ such that $\widetilde{K}_{ij}^{\text{PPR}} = \sqrt{K_{ij}^{\text{PPR}} K_{ji}^{\text{PPR}}}$ is symmetric and produces the same *logarithmic distance* (9) as K^{PPR}. Hence, the logarithmic similarity $\kappa(x, y) = \ln \widetilde{K}_{xy}^{\text{PPR}}$ is a proximity.

At the same time, the above logarithmic similarities are not kernels, as the corresponding matrices have negative eigenvalues.

This implies that being a proximity is not a stronger property than being a kernel. By Corollary 1, the square root of the distance induced by a proximity is also a distance. However, this square rooted distance need not generally be Euclidean, thus, Theorem 1 is not sufficient to conclude that the initial proximity is a kernel.

It can be verified that all logarithmic measures corresponding to the similarity measures under study preserve the natural order of distances $d(1, 2) < d(1, 3) < d(1, 4)$ for the example (5).

4   Numerical Comparison of Similarity Measures in the Context of Unsupervised Learning

Here we compare various kernels, proximities, and generally similarity measures in the context of unsupervised learning method – spectral clustering (for background on spectral clustering see, e.g., [19,42]). We test them on random undirected graphs that are built according to the stochastic block model.

More precisely, each graph $G = (V, E)$ has the following structure: it consists of two clusters $V = C_1 \cup C_2$ with the intracluster edge density p_{in} and the intercluster density p_{out}, i.e.

$$p_{\text{in}} = P\{(i, j) \in E \mid i, j \in C_1\} = P\{(i, j) \in E \mid i, j \in C_2\},$$
$$p_{\text{out}} = P\{(i, j) \in E \mid i \in C_1, j \in C_2\}.$$

We introduce the following reference classification:

$$\text{cls}_{\text{true}}[i] = k \text{ if } i \in C_k.$$

Given a similarity measure (matrix) K, which is computed using one of the basic graph matrices (W, L or P), we apply the spectral clustering algorithm to separate the graph into m clusters, $m = 2$ in our case. The algorithm we use is similar to the one proposed in [19]. Let us recap it here for completeness:

---

[5] On various alternative versions of the triangle inequality, we refer to [17].

find normalized eigenvectors of K that correspond to its largest m eigenvalues and put them into columns of matrix X; flip signs of column of X in a way that elements with maximum absolute values in each column are positive; run K-means algorithm on rows of X with m output clusters and place the result into the array cls.

We will measure the difference between two clusterings with m clusters on a set of n nodes by the following function:

$$\mathcal{E}(\mathrm{cls}_1, \mathrm{cls}_2) = 1 - \frac{1}{n} \max_{\sigma \in S_m} |\{i \in \{1, .., n\} | \sigma(\mathrm{cls}_1[i]) = \mathrm{cls}_2[i]\}|.$$

Here S_m is the group of all permutations on the set $\{1, 2, .., m\}$. This function corresponds to the minimum relative classification error that can be achieved by renumbering the clusters. Its computation is equivalent to solving the assignment problem of size m.

Since the transformation from W to K^{Katz}, K^{comm}, K^{df} is monotonic for eigenvalues and does not affect eigenvectors, these similarity measures all lead to the same result by spectral clustering. The similarity measures K^{heat} and K^{regL} are in the same sense equivalent to $-L = W - D$. $K^{\mathrm{n-heat}}$ is equivalent to $-\mathcal{L} = -D^{-1/2}LD^{-1/2}$ and to $D^{-1/2}WD^{-1/2}$. K^{PPR} and K^{heatPPR} are equivalent to $P = D^{-1}W$.

Hence, it is meaningful to test the clustering procedure on W, $-L$, $-\mathcal{L}$ and P. Looking ahead, we say that for the unbalanced case that we tested the typical error for W and $-L$ was about 0.4 that was much more than for other similarity measures. Hence, we included results only on P, $-\mathcal{L}$ and also added the results of spectral clustering algorithm from scikit-learn Python library, which is based on left eigenvectors of P.

The logarithmic transformation, however, changes both eigenvalues and eigenvectors. Hence, it is interesting to test it for different similarity measures. Since they all depend on some parameters, we minimize the error over the parameter space for each graph and than average it over the set of graph realizations.

4.1   Balanced Model

We tested the unsupervised learning algorithms on 100 graph realizations of 200 nodes stochastic block model with two clusters of 100 nodes each, intracluster density $p_{\mathrm{in}} = 0.1$ and intercluster density $p_{\mathrm{out}} = 0.02$. Errors, minimized over the parameter space and averaged over 100 graph realizations are shown in Fig. 1. The black thin bars correspond to the 95% confidence intervals. Spectral sklearn corresponds to the spectral clustering algorithm from scikit-learn Python library. Spectral P and Spectral NL correspond to the spectral clustering algorithm with $P = D^{-1}W$ and $I - \mathcal{L} = D^{-1/2}WD^{-1/2}$. The others correspond to the spectral clustering algorithm with logarithmic transformations of corresponding similarity measures. Black errorbars correspond to the 95% confidence interval.

We observe that all the tested methods provide roughly the same error that is around 0.01%. This is the manifestation of the fact that in the balanced case clustering is relatively easy.

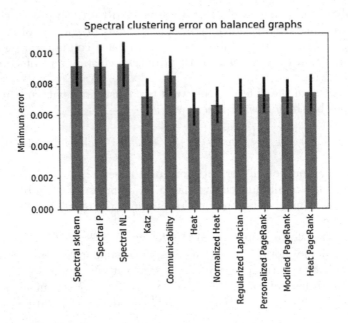

Fig. 1. Averaged minimum error for the balanced model.

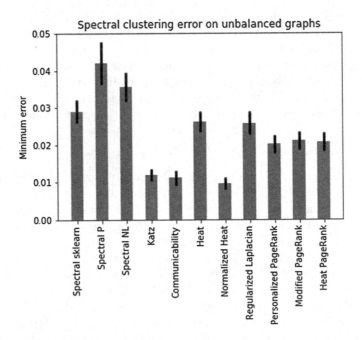

Fig. 2. Averaged minimum error for the unbalanced model.

4.2 Unbalanced Model

We also tested the algorithms on 1000 graphs of 200 nodes with two clusters of 50 and 150 nodes and the same edge densities $p_{in} = 0.1$ and $p_{out} = 0.02$. As expected, clustering unbalanced classes is more challenging.

Here we observe significant difference between results obtained with different similarity measures. Katz, communicability and normalized heat log measures lead to best results in this case.

We note that some other aspects of the comparative behavior of several kernels on graphs in clustering tasks have been studied in [26,30,40].

Acknowledgements. The work of KA and DR was supported by the joint Bell Labs Inria ADR "Network Science" and by UCA-JEDI Idex Grant "HGRAPHS", and the work of PC was supported by the Russian Science Foundation (project no.16-11-00063 granted to IRE RAS).

References

1. Avrachenkov, K., Mishenin, A., Gonçalves, P., Sokol, M.: Generalized optimization framework for graph-based semi-supervised learning. In: Proceedings of the 2012 SIAM International Conference on Data Mining, pp. 966–974 (2012)
2. Avrachenkov, K., Gonçalves, P., Sokol, M.: On the choice of kernel and labelled data in semi-supervised learning methods. In: Bonato, A., Mitzenmacher, M., Prałat, P. (eds.) WAW 2013. LNCS, vol. 8305, pp. 56–67. Springer, Cham (2013). doi:10.1007/978-3-319-03536-9_5
3. Avrachenkov, K., Chebotarev, P., Mishenin, A.: Semi-supervised learning with regularized Laplacian. Optim. Methods Softw. **32**(2), 222–236 (2017)
4. Avrachenkov, K., van der Hofstad, R., Sokol, M.: Personalized PageRank with node-dependent restart. In: Proceedings of International Workshop on Algorithms and Models for the Web-Graph, pp. 23–33 (2014)
5. Backstrom, L., Leskovec, J.: Supervised random walks: predicting and recommending links in social networks. Proc. ACM WSDM **2011**, 635–644 (2011)
6. Boley, D., Ranjan, G., Zhang, Z.L.: Commute times for a directed graph using an asymmetric Laplacian. Linear Algebra Appl. **435**(2), 224–242 (2011)
7. Chapelle, O., Schölkopf, B., Zien, A.: Semi-Supervised Learning. MIT Press, Cambridge (2006)
8. Chebotarev, P.: The graph bottleneck identity. Adv. Appl. Math. **47**(3), 403–413 (2011)
9. Chebotarev, P.: A class of graph-geodetic distances generalizing the shortest-path and the resistance distances. Discrete Appl. Math. **47**(3), 403–413 (2011)
10. Chebotarev, P.: The walk distances in graphs. Discrete Appl. Math. **160**(10–11), 1484–1500 (2012)
11. Chebotarev, P. Yu., Shamis, E.V.: On the proximity measure for graph vertices provided by the inverse Laplacian characteristic matrix. In: Abstracts of the conference "Linear Algebra and its Application", 10–12 June 1995, The Institute of Mathematics and its Applications, in conjunction with the Manchester Center for Computational Mathematics, Manchester, UK (pp. 6–7), URL http://www.ma.man.ac.uk/higham/laa95/abstracts.ps (1995)

12. Chebotarev, P.Y., Shamis, E.V.: The matrix-forest theorem and measuring relations in small social groups. Autom. Remote Control **58**(9), 1505–1514 (1997)
13. Chebotarev, P.Y., Shamis, E.V.: On a duality between metrics and Σ-proximities. Autom. Remote Control **59**(4), 608–612 (1998)
14. Chebotarev, P.Y., Shamis, E.V.: On proximity measures for graph vertices. Autom. Remote Control **59**(10), 1443–1459 (1998)
15. Chung, F.: Spectral graph theory, vol. 92. American Math. Soc. (1997)
16. Chung, F.: The heat kernel as the pagerank of a graph. Proc. Natl. Acad. Sci. **104**(50), 19735–19740 (2007)
17. Deza, M., Chebotarev, P.: Protometrics. arXiv preprint arXiv:1112.4829 (2011)
18. Dhillon, I.S., Fan, J., Guan, Y.: Efficient clustering of very large document collections. Data Min. sci. Eng. Appl. **2**, 357–381 (2001)
19. Dhillon, I.S., Guan, Y., Kulis, B.: Kernel k-means: spectral clustering and normalized cuts. Proc. ACM KDD **2004**, 551–556 (2004)
20. Estrada, E., Hatano, N.: Statistical-mechanical approach to subgraph centrality in complex networks. Chem. Phys. Lett. **439**, 247–251 (2007)
21. Estrada, E., Hatano, N.: Communicability in complex networks. Phys. Rev. E **77**(3), 036111 (2008)
22. Estrada, E., Silver, G.: Accounting for the role of long walks on networks via a new matrix function. J. Math. Anal. Appl. **449**, 1581–1600 (2017)
23. Fouss, F., Yen L., Pirotte, A., Saerens, M.: An experimental investigation of graph kernels on a collaborative recommendation task. In: Proceedings of the Sixth International Conference on Data Mining (ICDM 2006), pp. 863–868, IEEE (2006)
24. Fouss, F., Saerens, M., Shimbo, M.: Algorithms and Models for Network Data and Link Analysis. Cambridge University Press, Cambridge (2016)
25. Horn, R.A., Johnson, C.R.: Matrix Analysis, 2nd edn. Cambridge University Press, Cambridge (2013)
26. Ivashkin, V., Chebotarev, P.: Do logarithmic proximity measures outperform plain ones in graph clustering? In: Kalyagin, V.A., et al. (eds.) Models, Algorithms and Technologies for Network Analysis. Springer Proceedings in Mathematics & Statistics, vol. 197, pp. 87–105. Springer, Cham (2017)
27. Jacobsen, K., Tien, J.: A generalized inverse for graphs with absorption. arXiv preprint arXiv:1611.02233 (2016)
28. Katz, L.: A new status index derived from sociometric analysis. Psychometrika **18**(1), 39–43 (1953)
29. Kirkland, S.J., Neumann, M.: Group Inverses of M-matrices and Their Applications. CRC Press, Boca Raton (2012)
30. Kivimäki, I., Shimbo, M., Saerens, M.: Developments in the theory of randomized shortest paths with a comparison of graph node distances. Phys. A **393**, 600616 (2014)
31. Kondor, R.I., Lafferty, J.: Diffusion kernels on graphs and other discrete input spaces. In: Proceedings of ICML, pp. 315–322 (2002)
32. Lenart, C.: A generalized distance in graphs and centered partitions. SIAM J. Discrete Math. **11**(2), 293–304 (1998)
33. Liben-Nowell, D., Kleinberg, J.: The link-prediction problem for social networks. J. Assoc. Inform. Sci. Technol. **58**(7), 1019–1031 (2007)
34. Schoenberg, I.J.: Remarks to Maurice Fréchet's article "Sur la définition axiomatique d'une classe d'espace distanciés vectoriellement applicable sur l'espace de Hilbert". Ann. Math. **36**(3), 724–732 (1935)
35. Schoenberg, I.J.: Metric spaces and positive definite functions. Trans. Am. Math. Soc. **44**(3), 522–536 (1938)

36. Saerens, M.: Personal communication
37. Kandola, J., Shawe-Taylor, J., Cristianini, N.: Learning semantic similarity. In: Neural Information Processing Systems 15 (NIPS 2015). MIT Press (2002)
38. Shawe-Taylor, J., Cristianini, N.: Kernel Methods for Pattern Analysis. Cambridge University Press, Cambridge (2004)
39. Smola, A.J., Kondor, R.: Kernels and regularization on graphs. In: Learning Theory and Kernel Machines, pp. 144–158 (2003)
40. Sommer, F., Fouss, F., Saerens, M.: Comparison of graph node distances on clustering tasks. In: Villa, A.E.P., Masulli, P., Pons Rivero, A.J. (eds.) ICANN 2016. LNCS, vol. 9886, pp. 192–201. Springer, Cham (2016). doi:10.1007/978-3-319-44778-0_23
41. Vishwanathan, S.V.N., Schraudolph, N.N., Kondor, R., Borgwardt, K.M.: Graph kernels. J. Mach. Learn. Res. **11**, 1201–1242 (2010)
42. von Luxburg, U.: A tutorial on spectral clustering. Stat. Comput. **17**(4), 395–416 (2007)
43. Zhou, D., Schölkopf, B., Hofmann, T.: Semi-supervised learning on directed graphs. In: Proceeedings of NIPS, pp. 1633–1640 (2004)
44. Müller, K.-R., Mika, S., Rätsch, G., Tsuda, K., Schölkopf, B.: An Introduction to kernel-based learning algorithms. IEEE Trans. Neural Networks **12**(2), 181–202 (2001)

Endogenous Differentiation of Consumer Preferences Under Quality Uncertainty in a SPA Network

Bogumił Kamiński[1], Tomasz Olczak[1], and Paweł Prałat[2](✉)

[1] Decision Support and Analysis Unit, Warsaw School of Economics, Warsaw, Poland
[2] Department of Mathematics, Ryerson University, Toronto, ON M5B 2K3, Canada
pralat@ryerson.ca

Abstract. We study a duopoly market on which there is uncertainty of a product quality. Consumers adaptively learn about quality of products when they buy them (direct learning) or from other consumers with whom they are interacting in a social network modelled as a SPA graph (indirect learning). We show that quality uncertainty present in such a market leads to endogenous segmentation of consumers' preferences towards suppliers. Additionally, we show that in this setting, even if both companies have the same expected quality, the company with lower variance of quality will gain higher market share.

1 Introduction

In economics textbooks a discussion of network effects is usually limited to positive or negative externalities caused by preferential attachment process. In this paper we argue that below the surface of these evident and well known phenomena there is a layer of more subtle and less understood ones. Specifically the subject of our study is perception of product quality by consumers embedded in a network of social interactions induced by a SPA graph. We show how presence of quality uncertainty in this setting leads to diversification of quality expectations and endogenous differentiation of consumer preferences towards suppliers.

The significance of quality expectations for market mechanism has been originally revealed in the seminal paper of George Akerlof (1970). Using a market for second hand cars as an example, Akerlof has shown how presence of quality uncertainty leads to adverse selection of low quality products and degeneration or even failure of a market. More specifically, he considered a market comprised of two types of agents — car owners and potential car buyers. The first group was offering cars for sale, asking for prices reflecting their actual quality, but bids from the second group were based on the *average* quality of the cars on the market, as the true quality distribution was hidden from them. The mismatch of supply and demand discouraged owners of better than average cars causing them to withdraw from the market. This, in turn, triggered a feedback loop of gradual deterioration of quality of cars on sale until the market collapsed in the end.

© Springer International Publishing AG 2017
A. Bonato et al. (Eds.): WAW 2017, LNCS 10519, pp. 42–59, 2017.
DOI: 10.1007/978-3-319-67810-8_4

The discovery of this apparently simple mechanism of adverse selection has proven to be one of the most fruitful insights in modern economics, spawning development of vast literature and helping to explain market phenomena as diverse as drastic loss of value suffered by brand-new vehicles on their first days of use, difficulties of elderly people or young motorcyclists to get insurance cover, dearth of credit markets in underdeveloped countries or high unemployment among minorities. Its profound influence was eventually recognized by awarding Akerlof and his collaborators the Nobel Memorial Prize in Economic Sciences in 2001. With such prominence and after nearly half a century of research, one would hardly expect any novelty on the topic. So the paper of Izquierdo and Izquierdo (2007) came as a surprise by providing a new insight into the phenomena and suggesting that influence of uncertain quality on markets is still not entirely understood.

Akerlof (1970) and his followers considered *information asymmetry* between the transacting parties to be the necessary prerequisite for a market degradation to occur. This assumption has been considered so fundamental that the 2001 Nobel Prize was awarded to Akerlof, Spence and Stiglitz "for their analyses of markets with asymmetric information." Quite unexpectedly therefore Izquierdo and Izquierdo (2007) claimed that the same effect could be induced by quality uncertainty alone. To prove it they proposed an agent-based model in which consumers estimated quality of a product based upon experiences of their own and of their acquaintances. As they have demonstrated under the assumed adaptive quality estimation process the symmetry of supply and demand breaks down and market degenerates, down to a point of non-existence of equilibrium, in the same way as in the model of Akerlof (1970). The effect is more pronounced when there is only individual quality estimation, whereas social interaction mitigates it. In contrast to the model of Akerlof (1970) no a priori assumption of information asymmetry is required as it is replaced by endogenous differentiation of quality expectations in the population of consumers.

It may surprise at first that the mechanism described by Izquierdo and Izquierdo (2007) has not been revealed earlier during nearly half a century of research. However under scrutiny one will notice a fundamental difference in a way quality expectations are formed in the two models. Car buyers in Akerlof (1970) follow what is known as the rational expectations hypothesis (REH). In short REH assumes agents to know the "true" structural form of the underlying data generating process the parameters of which they estimate, and their subjective expectations to be consistent with this knowledge. So car buyers are assumed to have perfect knowledge of the average quality of cars on sale at any moment (although not of quality of each individual item). In contrast consumers of Izquierdo and Izquierdo (2007) follow the adaptive expectations hypothesis (AEH), which does not assume this accurate a priori knowledge. Instead, agents apply a simplistic first-order prediction error correction formula with exponentially decaying weights.

AEH and REH are the two extremes on the spectrum of expectation formulation methods. AEH as the earlier approach was commonly used in economics (for a classic example see Nerlove 1958) until the critique by Muth (1960) who shown its non-optimal statistical properties as the "backward-looking" biased estimator. Use of AEH was discouraged afterwards and replaced by REH as subsequently proposed by Muth (1961) and advocated by Lucas (1972) and Sargent (1973). The fact that REH has been integrated into the paradigm of the mainstream neoclassical economics does not discredit the adaptive approach nonetheless. In fact REH is often criticized for making too strong, psychologically unrealistic assumptions of agents rationality and their ability to perceive and process information (Evans and Honkapohja 2001). Another problem is that in many models REH results in multiple equilibria but does not indicate how to resolve the conundrum. For these reasons AEH is being actively researched as a viable alternative and many hybrid learning tactics combining AEH and REH are proposed (Frydman and Phelps 2013).

We hence find effects spotted by Izquierdo and Izquierdo (2007) as intriguing enough to deserve further exploration, although in a slightly modified setting. Both the discussed models required disgruntled agents, car sellers in Akerlof (1970) and consumers in Izquierdo and Izquierdo (2007), to retreat from the market for its contraction to occur. While this assumption could be justified in some circumstances, in many others it would be unrealistic. A person would rather look for substitutes than give up consumption altogether. Therefore in our model we assume a market with alternative suppliers of a homogeneous good and we allow consumers to switch suppliers to maximize satisfaction.

Note that although technically we consider a case of oligopoly, we are not interested in strategic interplay of suppliers. We use term "oligopoly" not in a classical sense but merely to signal that a key assumption of our model is ability of consumers to distinguish between multiple suppliers. The subject of our study is dynamics of consumer preferences under quality uncertainty and adaptive expectations. The ability of consumers to discriminate suppliers is a necessary prerequisite for it, but does not restrict the context to the textbook definition of oligopoly. Our conclusions are general and extend onto any market where consumers are able to perceive quality variability and discern suppliers, regardless of market power exercised by the latter. So they readily apply to monopolistic or perfect competition, as long as the key assumption of supplier distinction holds. In other words our aim is to study dynamics of quality expectations in isolation and we assume both supply and demand to be totally inelastic.

The remaining part of the paper is organized as follows. In Sect. 2 we describe the model of the market with quality uncertainty in detail. For better understating it will be presented in two variants for (a) finite and (b) continuous state space. In Sect. 3 we present results of the base analysis of an isolated consumer which do not take into account effects of social interaction. In Sect. 4 we extend the analysis with network effects induced by a socially realistic SPA connection graph. Section 5 presents the summary of the results and concludes.

2   Model

In this section, we define the model that we investigate. In order to validate robustness of presented results we present two variants of it. The first variant is a minimal specification that exhibits base properties we want to explore and better understand; it uses a finite state-space representation. The other variant has uncountable state-space and so it is more realistic, but obviously more challenging to handle.

2.1   Finite State-Space Model

Let $\mathcal{G} = (A, C)$ be a directed graph of connections between agents from set A. A connection $c \in C \subseteq A \times A$ is an ordered tuple $(a_1, a_2) \in C$ indicating that agent a_1 has influence on opinion of agent a_2. As mentioned, the graph is directed. It will be assumed at some point that $\mathcal{G}$ is a random geometric graph generated by the Spatial Preferential Attachment model that received some attention recently, see Sect. 4.1.

Suppose that there are two companies in the model, each providing some product or service. Customer buying product from company $i \in \{1, 2\}$ observes its value equal to a sample being random variable Q_i. It is assumed that samplings are independent. We take that Q_i is defined as an identity on the probability space $(\Omega, 2^{\Omega}, P_i)$, where $\Omega = \{-1, 0, 1\}$. We will say that -1 is a bad value of the product, 0 is a normal value of the product and 1 is an excellent value of the product. For this probability space we naturally define a probability mass function $p_i \colon \Omega \to]0, 1[$ (we assume that each state has strictly positive probability). In the model, we consider discrete time-steps. Each agent $a \in A$, in each time-step t, has evaluation of quality of company i as $e_{a,i}^t \in \{-1, 0, 1\}$, i.e. the agent believes that the company has bad, normal or excellent quality product. A vector of evaluations of agent a in time t is denoted as $\mathbf{e}_a^t$.

We start with time-step $t = 0$. Assume that $e_{a,i}^0$ is a random i.i.d. drawn from Q_i. The dynamics of the model is defined by the following procedure:

1. increase time-step $t \leftarrow t + 1$;
2. select $a \in A$ uniformly at random;
3. agent a selects company i to buy from as a draw from random variable $S(\mathbf{e}_a^t)$ specified below;
4. agent a observes quality of selected company i as q_i being a random sample from Q_i;
5. *individual learning*: agent a updates her beliefs $e_{a,i}^t$ as a draw from random variable $\Gamma(e_{a,i}^t, q_i)$ where Γ is specified below;
6. *social learning*: each agent b for which $(a, b) \in C$ updates her beliefs $e_{b,i}^t$ as a draw from random variable $\Delta(e_{a,i}^t, q_i)$, where Δ is specified below.

Definition of random variable S is the following:

– if $e_{a,i}^t = e_{a,j}^t$ then
$$\Pr(S = i) = \Pr(S = j) = 1/2;$$

– if $e^t_{a,i} > e^t_{a,j}$ then

$$\Pr(S = i) = \beta \text{ and } \Pr(S = j) = 1 - \beta, \text{ where } \beta \in [0.5, 1);$$

Definition of random variable Γ is the following:

$$\Pr(\Gamma = q_i) = \gamma \text{ and } \Pr(\Gamma = e^t_{a,i}) = 1 - \gamma, \text{ where } \gamma \in (0, 1).$$

Definition of random variable Δ is the following:

$$\Pr(\Delta = q_i) = \delta \text{ and } \Pr(\Gamma = e^t_{a,i}) = 1 - \delta, \text{ where } \delta \in [0, 1).$$

2.2   Continuous State Space

For simplicity, here we will discuss only the differences in specification of this model in comparison to the previous, finite state-space model. The model is analysed on the same graph $\mathcal{G}$ and has the same specification of dynamics. We provide new parameter names in this model, but there is a direct correspondence between these parameters and the parameters in finite state-space variant.

As before, we consider two companies. Customer buying a product from company i observes its value equal to sample from random variable $Q_i \sim \mathcal{N}(\mu_i, \sigma_i^2)$. The parameter pair (μ_i, σ_i) corresponds to pair $(p_i(-1), p_i(1))$ in the finite state space model.

We may assume that initially (that is, at time $t = 0$) agents have beliefs that are based on their first time purchases, i.e. $e^0_{i,a}$ is a sample from Q_i. Alternatively, we may assume that each agent start with a given value of $e^0_{i,a}$, say, $e^0_{i,a} = \mu_i$. The selection function $S(\mathbf{e}^t_a)$ is specified as:

$$S(\mathbf{e}^t_a) = \arg\max_i \{e^t_{i,a} + \varepsilon_i\},$$

where $\varepsilon_i \sim N(0, \omega^2)$ and are independent. Parameter ω corresponds to parameter β in finite state-space model.

After company i is selected agent a observes a sample of quality Q_i denoted as q_i.

Beliefs in the continuous state space model are deterministically updated. Individual learning is governed by the rule:

$$e^{t+1}_{i,a} \leftarrow (1 - \lambda_1)e^t_{i,a} + \lambda_1 q_i,$$

where $\lambda_1 \in (0, 0.5)$ is an individual learning parameter. Social learning follows the rule:

$$e^{t+1}_{i,b} \leftarrow (1 - \lambda_2)e^t_{i,b} + \lambda_2 q_i,$$

where $\lambda_2 \in [0, 0.5)$ is a social learning parameter. Observe that λ_1 corresponds to γ, and λ_2 to δ in the finite state-space model.

3   Baseline Analysis Without Network Effects

In this section, we investigate both models in a simple scenario where network does not affect the behaviour of agents (that is, $\delta = 0$ or $\lambda_2 = 0$, depending on the variant considered).

3.1   Finite State-Space Model

If $\delta = 0$, then our model can be thought of as a single customer model where an evolution of customer's state is given by a Markov process. The transition matrix $M_{9\times9}$ can be explicitly analytically derived.[1] Observe that the process is irreducible and positive recurrent, given the assumed domains of parameters. Therefore, it has a unique steady state π that is a solution of the following system of equations:

$$\begin{cases} \pi^T(M - I) = 0 \\ \pi^T 1 = 1 \end{cases}$$

Additionally, we can observe that then γ only influences the speed of convergence of the process as $M - I = \gamma X$, where X does not depend on γ. Under such observations π can be expressed as a function of five parameters: β, $p_1(-1)$, $p_1(1)$, $p_2(-1)$, and $p_2(1)$.

In order to assess the model we use the following metrics of the steady state:

- Mean evaluation of company i by the customer: $E(e_{a,i})$;
- Probability that company i is evaluated better than company j: $\Pr(e_{a,i} > e_{a,j})$;
- Probability that company i is evaluated equally to company j: $\Pr(e_{a,i} = e_{a,j})$.

Since the model is symmetric with respect to companies 1 and 2, in the following analysis we concentrate on company 1. In Figs. 1, 2 and 3 we show these metrics for the case when we assume that offers of companies are symmetric, i.e. $p_i(-1) = p_i(1)$, which means that $E(Q_i) = 0$ for both companies (the plots show averages over β uniformly distributed in intervals specified in subplot captions). We argue that this case is interesting because it represents the situation where both companies are equally good but only differ in the dispersion of their qualities. In this text we will solely concentrate on this scenario.

One can conclude from plots that in the steady state:

(F1) $\Pr(e_{a,i} = -1) > \Pr(e_{a,i} = 1)$ so $E(e_{a,i}) < 0$; on the average customers have negatively biased opinion; this bias is potentially significant and reaches ≈ -0.6, when the range of possible results is $[-1,1]$;

(F2) $p_i(1) < p_j(1) \Rightarrow E(e_{a,i}) > E(e_{a,j})$; company with higher variance has lower market share; this is the crucial finding of no-network model: it pays off to give customers service with predictable quality;

(F3) for high $\beta, p_1(1), p_2(1)$ we have bimodality, i.e. $\min_{i\in\{1,2\}} \Pr(e_{a,i} > e_{a,3-i}) > \Pr(e_{a,1} = e_{a,2})$; most likely client has a clear preference for one product or the other.

---

[1] We omit it in the text as it is large but easy to derive.

As a side note (not investigated in detail in this paper), let us observe that the system exhibits significantly nonlinear behaviour; shape of relationships of measured quantities changes with β.

The key question raised in this paper is how network structure affects the results F1–F3 (i.e., what happens when $\delta = 0$ is replaced by $\delta > 0$). In particular:

- how does the in-degree of a given agent influence bias of her preferences;
- do customers that are connected in a graph have correlated preferences;
- how does δ influence bias in evaluation of performance of companies;
- how does δ influence the presence of bimodality of preferences.

3.2   Continuous State-Space Model

We move to continuous state-space model but continue investigating the variant with no network effects, that is, when $\lambda_2 = 0$. We present two approaches to highlight tools that can be used in such situations. The first one is asymptotic in nature and provides a statement that holds asymptotically almost surely (and so can be considered to be more rigorous). The second one is based on simulations and can be applied for finite (but usually large) number of agents (and so can be considered to be more realistic).

Differential Equations Approach. As typical in random graph theory, all results in this subsection are asymptotic; that is, for $n = |A|$ tending to infinity. We say that an event holds asymptotically almost surely (a.a.s.) if it holds with probability tending to one as $n \to \infty$.

The general setting that is used in the differential equation method (Wormald 1999) is a sequence of random processes indexed by n (which in our case is

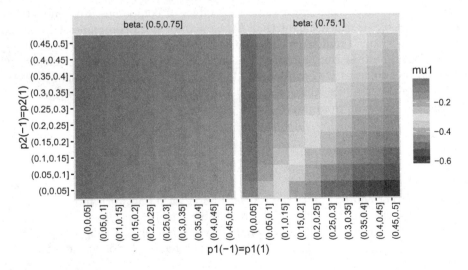

Fig. 1. $E(e_{a,1})$: mean of evaluation of company 1 for symmetric offers (mu1).

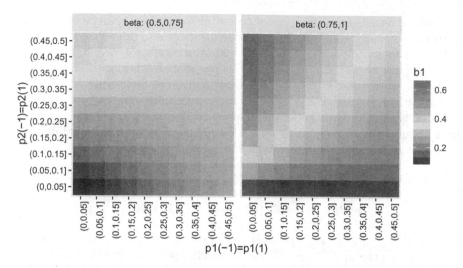

Fig. 2. $\Pr(e_{a,1} > e_{a,2})$: probability that company 1 has better evaluation than company 2 (b1).

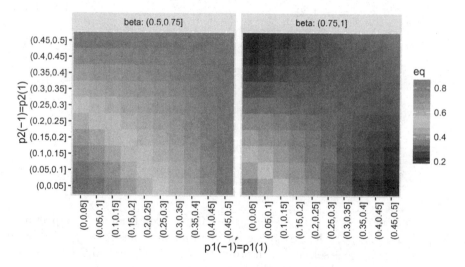

Fig. 3. $\Pr(e_{a,1} = e_{a,2})$: probability that companies are equally evaluated (eq).

the number of agents). The aim is to find asymptotic properties of the random process and the conclusion we aim for is that variables defined are well concentrated, which informally means that a.a.s. they are very close to certain deterministic functions. These functions arise as the solution to a system of ordinary first-order differential equations. One of the important features of this approach is that the computation of the approximate behavior of processes is clearly separated from the proof that the approximation is correct.

First, let us discretize the space of potential states agents can be in. Fix a real number $z > 0$, an integer k, and let us restrict ourselves to $(2k+1)$ values of possible believes for a company $c \in \{1, 2\}$: $\mu_c - zk/k, \mu_c - z(k-1)/k, \ldots, \mu_c, \ldots, \mu_c + z(k-1)/k, \mu_c + zk/k$. Each time some belief is updated, it is immediately rounded up or down to the nearest possible value (for a given company c).

For $-k \le i, j \le k$, let $X_{i,j}(t)$ be a random variable counting the number of agents of type (i, j) with belief about product 1 equal to $\mu_1 + zi/k$ and with belief about product 2 equal to $\mu_2 + zj/k$. Let $q(i, j)$ be the probability that agent of type (i, j) buys product 1; that is,

$$q(i, j) = P\left(\mu_1 + zi/k + N(0, \omega) \ge \mu_2 + zj/k + N(0, \omega)\right)$$
$$= P\left(N(0, 2\omega) \ge (\mu_2 - \mu_1) + z(j - i)/k\right).$$

The probability that she buys product 2 is, of course, $q(j, i) = 1 - q(i, j)$. Now, let $r(s, i, \mu, \sigma^2)$ be the probability that an agent changes her believes from $\mu + zs/k$ to $\mu + zi/k$ after buying product with the corresponding quality distribution $N(\mu, \sigma^2)$ (and after rounding, of course); that is, for $-k < i < k$

$$r(s, i, \mu, \sigma^2) = P\left(\mu + \frac{z(i - 1/2)}{k} \le (1 - \lambda_1)\left(\mu + \frac{zs}{k}\right) + \lambda_1 N(\mu, \sigma^2) \le \mu + \frac{z(i + 1/2)}{k}\right)$$
$$= P\left(\frac{z(i - s(1 - \lambda_1) - 1/2)}{k\lambda_1} \le N(\mu, \sigma^2) - \mu \le \frac{z(i - s(1 - \lambda_1) + 1/2)}{k\lambda_1}\right)$$
$$= P\left(\frac{z(i - s(1 - \lambda_1) - 1/2)}{k\lambda_1} \le N(0, \sigma^2) \le \frac{z(i - s(1 - \lambda_1) + 1/2)}{k\lambda_1}\right).$$

For the two extreme values ($i = -k$ and $i = k$) we have

$$r(s, -k, \mu, \sigma^2) = P\left(N(0, \sigma^2) \le \frac{z(i - s(1 - \lambda_1) + 1/2)}{k\lambda_1}\right)$$
$$r(s, k, \mu, \sigma^2) = P\left(\frac{z(i - s(1 - \lambda_1) - 1/2)}{k\lambda_1} \le N(0, \sigma^2)\right).$$

Our goal is to estimate the conditional expectation $\mathbb{E}\left[X_{i,j}(t + 1) - X_{i,j}(t) \mid \mathbf{X}\right]$ (given the set $\mathbf{X}$ of all variables $X_{i,j}(t)$). Note that an agent of type (i, j) is selected with probability $X_{i,j}(t)/n$. Conditioning on this event, the probability she stays within this group is equal to

$$q(i, j) \cdot r(i, i, \mu_1, \sigma_1^2) + q(j, i) \cdot r(j, j, \mu_2, \sigma_2^2).$$

For $s \ne i$, an agent of type (s, j) is selected with probability $X_{s,j}(t)/n$, and conditioning on that, she becomes of type (i, j) with probability

$$q(s, j) \cdot r(s, i, \mu_1, \sigma_1^2).$$

Similarly, for $y \ne j$, an agent of type (i, y) is selected with probability $X_{i,y}(t)/n$, and conditioning on that, she becomes of type (i, j) with probability

$$q(y, i) \cdot r(y, j, \mu_2, \sigma_2^2).$$

It follows that

$$\mathbb{E}\left[X_{i,j}(t+1) - X_{i,j}(t) \mid \mathbf{X}\right] = -\frac{X_{i,j}(t)}{n}$$

$$+ \sum_{s=-k}^{k} \frac{X_{s,j}(t)}{n} q(s,j) r(s,i,\mu_1,\sigma_1^2)$$

$$+ \sum_{y=-k}^{k} \frac{X_{i,y}(t)}{n} q(y,i) r(y,j,\mu_2,\sigma_2^2).$$

For simplicity, as this is just an illustration of the general method, we may assume that, say, $X_{0,0} = n$, and other values are 0 (that is, initially every agent believes that product 1 has quality μ_1 and product 2 has quality μ_2). Any other scenario can be investigated the same way affecting only the initial value for the system of differential equations we are about to set up.

Now, one can scale everything down (both the time n and the number of members of each group, n) to get the system of differential equations. Here, function $f_{i,j}(x)$ is used to model random variable $X_{i,j}(xn)/n$. We get the system of $(2k+1)^2$ equations: for $-k \le i,j \le k$,

$$f'_{i,j}(x) = -f_{i,j}(x)$$

$$+ \sum_{s=-k}^{k} f_{s,j}(x) q(s,j) r(s,i,\mu_1,\sigma_1^2)$$

$$+ \sum_{y=-k}^{k} f_{i,y}(x) q(y,i) r(y,j,\mu_2,\sigma_2^2),$$

with the initial value $f_{0,0}(0) = 1$ and $f_{i,j}(0) = 0$ if $|i| + |j| > 0$.

Finally, the differential equations method (introduced and developed by Wormald 1999) can be used to show that our random variables are well-concentrated around their expectations. Using the general purpose theorem (Theorem 5.1 in Wormald 1999), we get that a.a.s. for any $-k \le i,j \le k$, and any t, we have

$$X_{i,j}(t) = (1 + o(1)) f_{i,j}(t/n) n.$$

In order for the discretized model to approximate with good accuracy the original, continuous model, one should take: (i) z large enough so that, for any company $c \in \{1,2\}$, in the original model, the number of agents that get belief below $\mu_c - z$ or above $\mu_c + z$ is negligible; (ii) k large enough to capture a large spectrum of beliefs. As a result, plotting all $(2k+1)^2$ functions is an impossible task but the following three functions should describe well the behaviour of the system:

$$f_=(x) := \sum_{i=-k}^{k} f_{i,i}(x) \text{ (fraction of agents that equally like both products)}$$

$$f_>(x) := \sum_{i=-k}^{k} \sum_{j=-k}^{i-1} f_{i,j}(x) \text{ (fraction of agents that like product 1 more)}$$

$$f_<(x) := \sum_{i=-k}^{k} \sum_{j=i+1}^{k} f_{i,j}(x) \text{ (fraction of agents that like product 2 more)}$$

Figure 4 presents an example of the dynamics of the process. If $\sigma_2 > \sigma_1$, then the company with a product having lower variability gains higher market share. Interestingly, the transient behavior of this simulation is that initially company 2 gains market share, but then starts losing it as $f_=$ drops and $f_>$ continues to increase. As k increases $f_=$ will tend to 0 in general. Economically this might have twofold implications: (1) it might be profitable to launch even a product that is known to be inferior than competition because profits that can be reaped in the initial period might justify it and (2) investors should look at initial success of a product with care as it might be only transient characteristic of a system.

Finally, let us stress again that we present an application of the differential equations method in a very simple setting but it can be easily generalized to

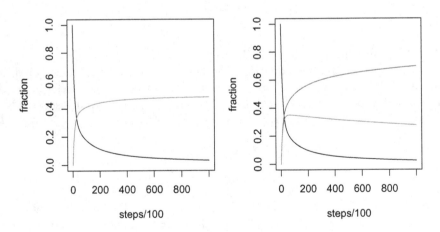

Fig. 4. Example of the dynamics of the model. Black line represents $f_=$, red line $f_>$, and green line $f_<$. On the left plot we consider a symmetric case: $\sigma_1 = \sigma_2 = 1$ (red and green lines overlap). On the right plot an asymmetric case is considered: $\sigma_1 = 1$ and $\sigma_2 = 1.25$. In both plots $\mu_1 = \mu_2$, $\lambda_1 = 0.5$ and $\omega = 0.1$. In approximation we used $z = 5$ and $k = 100$. Initial beliefs are equal to μ_c, for both companies. (Color figure online)

more sophisticated scenarios. For example, when agents are not selected uniformly at random from set A (see step 2) but, instead, with probability that is affected by type (i, j) a given agent is of. Or perhaps agents select company to buy from (see step 3) with probability that depends on how many other agents have similar believes. In these examples, this method seems to be the only tool one can use. On the other hand, in our example, one can avoid using it as it is straightforward to predict (a.a.a., as always) how many agents at time xn (for some constant x) made ℓ purchases ($\ell = 0, 1, \dots$). Then, each agent can be investigated independently (based on the assigned value of ℓ) using Markov processes, as for the finite state-space model (but with $(2k + 1)$ states instead of 9). Putting things together, we can calculate the expected number of agents of a given type and the concentration will follow from standard tools (such as Chernoff bound), since the corresponding events are independent.

Simulation Approach. The simulation was run for 1'000'000 iterations which was enough for it to reach steady-state. In Figs. 5 and 6 we can observe that expected evaluation of company 1 is negative and that it decreases with σ_1 and increases with σ_2. Additionally increase of ω reduces those differences (as customers behave more randomly).

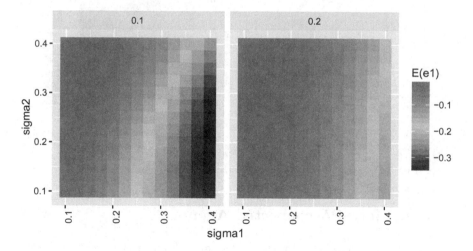

Fig. 5. $E(e_{a,1})$: mean evaluation of company 1 (for two values of ω).

In Fig. 7 we show distribution of beliefs of agents for $\sigma_1 = \sigma_2 = 1$, $\lambda_1 = 0.1$ and $\lambda_2 = 0$. For this parameterization we have that $E(e_{a,i}) \approx -0.27$ and correlation between $e_{a,1}$ and $e_{a,2}$ equals to approximately -0.65. The crucial thing is that as depicted on the plot we observe bimodality in the beliefs of agents — both in one belief and for combination of two beliefs.

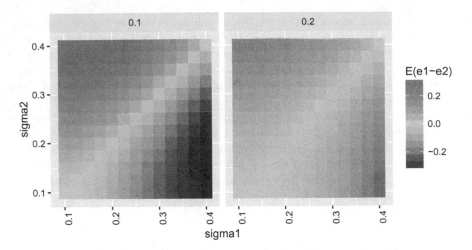

Fig. 6. $E(e_{a,1} - e_{a_2})$: mean difference of evaluation of company 1 and 2 (for two values 0.1 and 0.2 of ω).

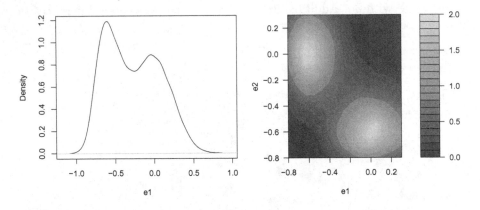

Fig. 7. Density of $e_{a,1}$ and joint density distribution of $(e_{a,1}, e_{a,2})$.

4   Results for SPA-connected Agents

In this section we extend the analysis onto network effects induced by the SPA model that is a stochastic and geometric model of complex networks. Here it is used to model social connections between agents. We first start with specification of SPA model and then present analysis of model performance.

4.1   Spatial Preferential Attachment Model

The *Spatial Preferential Attachment* (SPA) model, introduced in Aiello et al. (2009), is designed as a model for the World Wide Web and combines geometry and preferential attachment, as its name suggests. Setting the SPA model apart is the incorporation of "spheres of influence" to accomplish preferential attachment: the greater the degree of a vertex, the larger its sphere of influence, and hence the higher the likelihood of the vertex gaining more neighbours.

We now give a precise description of the SPA model. Let $S = [0,1]^m$ be the unit hypercube in $\mathbb{R}^m$, equipped with the torus metric derived from any of the L_k norms. This means that for any two points x and y in S,

$$d(x,y) = \min\left\{||x - y + u||_k : u \in \{-1,0,1\}^m\right\}.$$

The torus metric thus "wraps around" the boundaries of the unit square; this metric was chosen to eliminate boundary effects. The parameters of the model consist of the *link probability* $p \in [0,1]$, and two positive constants A_1 and A_2, which, in order to avoid the resulting graph becoming too dense, must be chosen so that $pA_1 < 1$. The SPA model generates stochastic sequences of directed graphs $(G_t : t \geq 0)$, where $G_t = (V_t, E_t)$, and $V_t \subseteq S$. Let $\deg^-(v,t)$ be the in-degree of the vertex v in G_t, and $\deg^+(v,t)$ its out-degree. We define the *sphere of influence* $S(v,t)$ of the vertex v at time $t \geq 1$ to be the ball centered at v with volume $|S(v,t)|$ defined as follows:

$$|S(v,t)| = \min\left\{\frac{A_1\deg^-(v,t) + A_2}{t}, 1\right\}. \tag{1}$$

The process begins at $t = 0$, with G_0 being the null graph. Time step $t, t \geq 1$, is defined to be the transition between G_{t-1} and G_t. At the beginning of each time step t, a new vertex v_t is chosen *uniformly at random* from S, and added to V_{t-1} to create V_t. Next, independently, for each vertex $u \in V_{t-1}$ such that $v_t \in S(u, t-1)$, a directed link (v_t, u) is created with probability p. Thus, the probability that a link (v_t, u) is added in time-step t equals $p\,|S(u, t-1)|$.

The SPA model produces scale-free networks, which exhibit many of the characteristics of real-life networks (see Aiello et al. 2009, Cooper et al. 2014). In Janssen et al. (2013a), it was shown that the SPA model gave the best fit, in terms of graph structure, for a series of social networks derived from Facebook. In Janssen et al. (2013b), some properties of common neighbours were used to explore the underlying geometry of the SPA model and quantify vertex similarity based on distance in the space. However, the distribution of vertices in space was assumed to be uniform Janssen et al. (2013b) and so in Janssen et al. (2016) non-uniform distributions were investigated which is clearly a more realistic setting. Finally, in Ostroumova Prokhorenkova et al. 2017 modularity of this model was investigated, which is a global criterion to define communities and a way to measure the presence of community structure in a network.

Specifically, in Aiello et al. (2009) (Theorem 1.1) it was proved that the SPA model generates a graph with a power law in-degree distribution with exponent $1 + 1/(pA_1)$. On the other hand, the average out-degree is asymptotic to $pA_2/(1 - pA_1)$ (see Theorem 1.3 in Aiello et al. 2009). In this text we take $m = 2$, $k = 2$ (two-dimensional Euclidean space), and a graph of $|A| = 10,000$ agents, $A_1 = 1$, $A_2 = 6$ and $p = 0.5$. This means that in our simulation power law has coefficient 3 and the average out-degree (and so also in-degree) is 6.

4.2   Results: Finite State-Space Model

The model is still Markov process, however its state space has now size $9^{|A|}$. Moreover, the process is still irreducible and positive recurrent given the assumed domains of parameters. Therefore, it has a unique steady state π.

Table 1. Influence of variables on target characteristics approximated by linear regression; experiment setup: $\gamma \in \{0.25, 0.5\}$, $\delta \in \{0, 0.25, 0.5\}$, $\beta \in \{0.6, 0.7, 0.8, 0.9\}$ and $p_i(-1) = p_i(1) \in \{0.1, 0.2, 0.3, 0.4\}$. Estimates significant at 0.001 marked with *.

Variable	Mean	$p_1(-1) = p_1(1)$	$p_2(-1) = p_2(1)$	β	γ	δ
degcor1	0.0583	0.0888*	−0.0875*	0.2584*	−0.0036	0.1723*
$E(e_{a,1})$	−0.1346	−0.4789*	0.2230*	−0.5920*	−0.0495	0.2300*
$\Pr(e_{a,1} > e_{a,2})$	0.3219	−0.1408*	0.5299*	0.0735*	0.0035	−0.0201
$\Pr(e_{a,1} = e_{a,2})$	0.3566	−0.3869*	−0.3878*	−0.1352*	−0.0050	0.0450
edgecor1	0.0594	0.0151	−0.0131	0.0245*	0.0712*	0.2064*

In the following analysis by degcor1 we denote correlation of in-degree of agent a and her $e_{a,1}$ and by edgecor1 we denote correlation of $e_{a,1}$ and $e_{b,1}$ for all agents a and b that are connected by an edge.

In Table 1 we concentrate our analysis on means and want to understand the influence of parameter δ on the results. The analyzed data were collected from 384 runs of the simulation for Cartesian product of $p_1, p_2 \in \{0.1, 0.2, 0.3, 0.4\}$, $\beta \in \{0.6, 0.7, 0.8, 0.9\}$, $\gamma \in \{0.25, 0.5\}$ and $\delta \in \{0, 0.25, 0.5\}$. We report the parameters of the influence of input variables on simulation outputs estimated using linear regression metamodel.

On the average, agents with higher in-degrees have higher evaluations of product qualities. Remembering that it is on the average negative it means that higher in-degree reduces bias in evaluation of product quality. Also we observe that agents that are connected by edge have on the average positive correlation of opinions. The analysis of parameters at δ variable shows that it has relatively low impact of the structure of preferences in the population ($\Pr(e_{a,1} > e_{a,2})$ and $\Pr(e_{a,1} = e_{a,2})$ variables) — the structure of bimodality is approximately similar to no-network case. However, higher values of δ strongly reduce bias of estimates ($E(e_{a,1})$) and increase correlation of opinion with degree and between agents that are connected in the network.

4.3   Results: Continuous State-Space Model

The results are analogous to the finite state-space case. In Table 2 we still concentrate our analysis on means but this time we want to observe the influence of parameter λ_2 on the results. The analyzed data were collected from 384 runs of the simulation for Cartesian product of $\sigma_1, \sigma_2 \in \{0.1, 0.2, 0.3, 0.4\}$, $\omega \in \{0.6, 0.7, 0.8, 0.9\}$, $\lambda_1 \in \{0.25, 0.5\}$ and $\lambda_2 \in \{0, 0.25, 0.5\}$. We report the

Table 2. Influence of variables on target characteristics approximated by linear regression; experiment setup: $\lambda_1 \in \{0.25, 0.5\}$, $\lambda_2 \in \{0, 0.25, 0.5\}$, $\omega, \sigma_1, \sigma_2 \in \{0.1, 0.2, 0.3, 0.4\}$. Estimates significant at 0.001 marked with *.

Variable	Mean	σ_1	σ_2	ω	λ_1	λ_2
degcor1	0.0731*	0.3940*	−0.1352*	−0.3431*	0.1650*	0.2022*
$E(e_{a,1})$	−0.0458*	−0.3502*	0.0873*	0.2460*	−0.1357*	0.0536*
$\Pr(e_{a,1} > e_{a,2})$	0.5010*	−0.6479*	0.6559	0.0064	−0.0045	0.0038
edgecor1	0.2027	0.0190	−0.0187	−0.0353	0.0620	0.5657*

parameters of the influence of input variables on simulation outputs estimated using linear regression metamodel.

On the average, agents with higher in-degree have higher evaluation of product quality. Remembering that it is on the average negative it means that higher in-degree reduces bias in evaluation of product quality. Also we observe that agents that are connected by edge have on the average positive correlation of opinions. The analysis of parameters at λ_2 variable shows that it has relatively low impact of the structure of preferences in the population ($\Pr(e_{a,1} > e_{a,2})$ variable) — the structure of bimodality is approximately similar to no-network case. However, higher λ_2 reduces bias of estimates ($E(e_{a,1})$) and increases correlation of opinion with degree and between agents that have a connection in a graph.

5   Concluding Remarks

Influence of quality uncertainty on markets has been traditionally studied in context of asymmetric information and rational expectations, an approach rooted in the seminal publication of Akerlof (1970). In this setting, uncertain quality causes a market to degenerate or even vanish altogether. Izquierdo and Izquierdo (2007) have demonstrated that equivalent results are obtained on markets without an a priori assumption of information asymmetry but with consumers having adaptive expectations of quality. In this paper we contributed to this stream of research by proposing a model of a market with multiple suppliers and consumers adaptively switching suppliers to maximize satisfaction. We made several interesting observations.

First, we noticed that under assumption of adaptive expectations quality uncertainty is a sufficient condition for endogenous differentiation of consumer preferences towards suppliers. As depicted on the joint density plot on Fig. 7 this effect takes shape of a bimodal distribution of expected quality with majority of consumers having clear preference for one of the suppliers and almost none being indifferent. Interestingly the effect is observed regardless if network effects are taken into account or not. Traditionally in economic modelling this kind of horizontal differentiation of preferences is attributed to exogenous factors such as purposeful diversification of product characteristics by sellers to increase their market power. As we have shown similar effects may occur spontaneously and without intentional effort, by means of random variations of a product quality.

Next, we found out that lowering quality variability provides a competitive advantage to suppliers, as those who second-order dominate quality distribution of the competitors, systematically increase their market share to eventually take over the market (Fig. 4). Note that we did not assume consumers to be risk averse so this effect emerged as the endogenous property of the model. This finding may have important practical implications as it provides credible justification for implementing quality assurance policies such as TQM or Six Sigma which are sometimes criticised for being merely costly "fads" having no theoretical underpinning (Linderman et al. 2003; Schroeder et al. 2008).

Finally by embedding consumers in a socially realistic SPA network we have the following findings: (a) agents with higher in-degree are better informed, (b) there is a correlation of beliefs of agents that are connected in a network, (c) higher rate of learning from neighbours reduces average bias of expectations. The analysis of dynamics of this process shows that the time to reach steady-state in the model is dramatically accelerated by social interaction as more signals are reaching the customers per one time period. This discovery reinforces the above practical conclusions regarding quality assurance policies as it indicates that the observed effects may strongly influence real markets, where information about bad quality product can spread fast and be hard to erase later. This effect is explained by our model and has been witnessed many times by companies in social media like Facebook or Twitter.

In next steps we will test robustness of our results under more "rational" Bayesian quality estimators. Another interesting question to be addressed is if lower quality could be compensated by its lower volatility i.e. is there a trade-off between expected quality and its variance. Numerical experiments confirm such possibility but this remains to be proven.

References

Aiello, W., Bonato, A., Cooper, C., Janssen, J., Prałat, P.: A spatial web graph model with local influence regions. Internet Math. **5**, 175–196 (2009)

Akerlof, G.A.: The market for "lemons": quality uncertainty and the market mechanism. Q. J. Econ. **84**(3), 488–500 (1970)

Cooper, C., Frieze, A., Prałat, P.: Some typical properties of the spatial preferred attachment model. Internet Math. **10**, 27–47 (2014)

Evans, G.W., Honkapohja, S.: Learning and Expectations in Macroeconomics. Princeton University Press (2001)

Frydman, R., Phelps, E.S. (eds.): The Way Forward for Macroeconomics. Princeton University Press (2013)

Izquierdo, S.S., Izquierdo, L.R.: The impact of quality uncertainty without asymmetric information on market efficiency. J. Bus. Res. **60**(8), 858–867 (2007)

Janssen, J., Hurshman, M., Kalyaniwalla, N.: Model selection for social networks using graphlets. Internet Math. **8**(4), 338–363 (2013a)

Janssen, J., Prałat, P., Wilson, R.: Geometric graph properties of the spatial preferred attachment model. Adv. Appl. Math. **50**, 243–267 (2013b)

Janssen, J., Prałat, P., Wilson, R.: Non-uniform distribution of nodes in the spatial preferential attachment model. Internet Math. **12**(1–2), 121–144 (2016)

Linderman, K., Schroeder, R.G., Zaheer, S., Choo, A.S.: Six sigma: a goal-theoretic perspective. J. Oper. Manage. **21**(2), 193–203 (2003)

Lucas, R.E.: Expectations and the neutrality of money. J. Econ. Theory **4**(2), 103–124 (1972)

Muth, J.F.: Optimal properties of exponentially weighted forecasts. J. Am. Statist. Assoc. **55**(290), 299–306 (1960)

Muth, J.F.: Rational expectations and the theory of price movements. Econometrica J. Econ. Soc. 315–335 (1961)

Nerlove, M.: Adaptive expectations and cobweb phenomena. Q. J. Econ. **72**(2), 227–240 (1958)

Ostroumova Prokhorenkova, L., Prałat, P., Raigorodskii, A.: Modularity of complex networks models. In: Bonato, A., Graham, F.C., Prałat, P. (eds.) WAW 2016. LNCS, vol. 10088, pp. 115–126. Springer, Cham (2016). doi:10.1007/978-3-319-49787-7_10

Sargent, T.J.: Rational expectations, the real rate of interest, and the natural rate of unemployment. Brookings Papers Econ. Activity **2**, 429–480 (1973)

Schroeder, R.G., Linderman, K., Liedtke, C., Choo, A.S.: Six sigma: definition and underlying theory. J. Oper. Manage. **26**(4), 536–554 (2008)

Wormald, N.: The Differential Equation Method for Random Graph Processes and Greedy Algorithms. Lectures on Approximation and Randomized Algorithms, pp. 73–155 (1999)

High Degree Vertices and Spread of Infections in Spatially Modelled Social Networks

Joshua Feldman and Jeannette Janssen$^{(\boxtimes)}$

Department of Mathematics and Statistics, Dalhousie University,
Halifax, NS, Canada
{Joshua.Feldman,Jeannette.Janssen}@dal.ca

Abstract. We examine how the behaviour of high degree vertices in a network affects whether an infection spreads through communities or jumps between them. We study two stochastic susceptible-infected-recovered (SIR) processes and represent our network with a spatial preferential attachment (SPA) network. In one of the two epidemic scenarios we adjust the contagiousness of high degree vertices so that they are less contagious. We show that, for this scenario, the infection travels through communities rather than jumps between them. We conjecture that this is not the case in the other scenario, when contagion is independent of the degree of the originating vertex. Our theoretical results and conjecture are supported by simulations.

Keywords: Spatial graph model · Preferential attachment · Infection in networks · Contact process

1 Introduction

While community structure plays an important role in the spread of infections [14], there are few analytic results on the topic and it is unclear precisely how clustering interacts with other network properties. Part of the difficulties in this area stem from the notion that communities consist of disjoint groups or small cycles. Recently, however, many have taken an approach to studying community structure by embedding vertices in a metric space [1,4,6,7,12]. One can interpret the metric space as a feature space in which nearby vertices have more affinity than the vertices at a distance and, accordingly, close vertices have a higher probability of being connected. Since a community is a group of individuals who share some similarities, we represent communities as geometric clusters. The use of spatial networks allows for a more nuanced notion of community where groups can overlap and boundaries are fuzzy. Not only is this approach more realistic, but it is also easier to analyze. We will exploit the mathematical tractability of a spatial model to study the interaction between community structure and the spread of infections.

Specifically, our focus will be how the behaviour of high degree nodes changes whether an infection spreads through communities or jumps between them.

© Springer International Publishing AG 2017
A. Bonato et al. (Eds.): WAW 2017, LNCS 10519, pp. 60–74, 2017.
DOI: 10.1007/978-3-319-67810-8_5

This question is important because understanding who spreads diseases between communities can help guide interventions. For example, as [14] show, vaccinating hosts who bridge communities can be more effective than vaccinating highly connected individuals. If high degree vertices connect communities, then these two strategies amount to the same thing.

We model our infection with a stochastic susceptible-infected-recovered (SIR) process in discrete time. Susceptible vertices can be infected in the future, infected vertices are currently sick, and recovered vertices have gained immunity from a previous infection. To control the behaviour of high degree vertices, we recognize that infections spread through contacts (i.e. sexual contact, airborn contact, etc.), but a network edge only refers to the *potential* for contacts. [13] demonstrate that the number of contacts made with a neighbour has a significant effect on epidemic dynamics. We consider two scenarios in which the "popular" vertices behave differently: (A) high degree and low degree vertices make the same number of total contacts per time step, so highly connected vertices make fewer contacts with any individual neighbour and (B) the time vertices spend with all their neighbours per time step is proportionate to their degree, so each vertex gets an equal amount of time with any individual neighbours. In scenario A, high degree nodes have many weak connections and, in scenario B, high degree and low degree vertices have connections of equal strength so high degree vertices should pass on the disease to more neighbours. We note that contacts are not reciprocal—two vertices can make a different number of contacts with one another. To model the relationship between contacts per neighbour per time step and the probability of infecting a susceptible neighbour in a time step, we use an STI model developed by Garnett and Anderson in [5].

To model our network, we use the Spatial Preferential Attachment (SPA) Model proposed in [1]. These networks are sparse power-law graphs with positive clustering coefficients [1]. It has been shown that the SPA model fits real-life social networks well [8]. Since we are working with a stochastic process on a random network, we modify the SPA model to remove some randomness. We show that, compared to the SPA model, vertices in our modified networks have the same expected degree and the overall degree distribution has the same power-law coefficient.

Using techniques developed by [9], we will show that when popular high degree nodes are less infectious (scenario A), the infection will travel slowly through the metric space and respect the community structure. We also conjecture that in scenario B, the infection will take long jumps between communities. To support our result and conjecture, we run simulations on the modified SPA model. While there are numerical studies exploring the relationship between the behaviour of highly connected individuals and the spread of disease with respect to community structure [14], we believe these are the first analytic results on the topic.

Our research presents new strategies for understanding communities in networks. Networks generated by the modified SPA model display many properties of real-world systems, and are more tractable than those generated by the original SPA model. More generally, by representing communities with a continuous feature space rather than with disjoint subsets, we can easily leverage geometric properties to prove otherwise difficult results. The work presented here develops techniques for understanding this geometric conception of community structure in networks.

2   Definitions

Here we present the definitions of the SPA model, the modified SPA model, and our model of the infection.

2.1   The SPA Model

The SPA model was first proposed in [1]. It is a spatial digraph model where vertices are embedded in a metric space. The metric space represents the feature space, which reflects the attributes of the vertices that determine their linking patterns. The model indirectly incorporates the principle of *preferential attachment*, first proposed by Barabasi and Albert (BA) [2], through the notion of *spheres of influence* around every vertex.

Vertices with a larger in-degree have a sphere of influence with greater volume, but as time progresses the spheres of influence of all nodes decrease. In the BA model, the preferential attachment came from a probability of a newcomer connecting to the old vertices. In the SPA model, we use the sphere of influence to incorporate the preferential attachment process. If a newcomer falls within an older vertex's sphere of influence, we connect them.

Specifically, vertices are embedded in a hypercube C of dimension d with unit volume. We endow the hypercube with the torus metric derived from any of the L_p norms. The torus metric is used to avoid edge effects. If x and y are two vertices in C, the distance between them is given by:

$$d(x, y) = \min\{\|x - y + u\|_p : u \in \{-1, 0, 1\}^m\}$$

The SPA model has parameters $A_1 \in (0, 1)$ and $A_2 \in [0, \infty)$. (The original SPA model also has a parameter p representing the conditional probability that a vertex which falls into the sphere of influence of vertex, actually links to that vertex. We here assume this to be 1.)

The model consists of a stochastic sequence of n graphs $\{G_t = (V_t, E_t)\}_{0 \le t \le n}$ with $V_t \subset C$. The index t is interpreted as the t-th time step. At each time t, the sphere of influence $S(v, t)$ of a vertex $v \in V_t$ is the ball centered at v with volume

$$|S(v,t)| = \min\left\{ \frac{A_1 \deg^-(v,t) + A_2}{t}, 1 \right\} \tag{1}$$

Let G_0 be the null graph. Given G_{t-1}, we define $G_t = (V_t, E_t)$ as follows. $V_t = V_{t-1} \cup \{v_t\}$ where v_t is placed uniformly at random in C. The edge set $E_t = E_{t-1} \cup \{(v_t, u) \mid v_t \in S(u, t)\}$.

We now review the relevant properties of the SPA model. As shown in [1], the network has a power law degree distribution, with an exponent of $1 + \frac{1}{A_1}$.

The geometric nature of the network implies that there is a high amount of local clustering [10]. In [3], logarithmic bounds on the directed diameter were given. In [9] it was shown that the effective undirected diameter is also logarithmically bounded.

2.2 The Modified SPA Model

While the SPA model generates spatial graphs that fit empirical networks well, we must modify the model to make it mathematically tractable for our purposes. Since the behaviour of a vertex during its early life significantly determines its late time behaviour, working with the preferential attachment model can be difficult. Instead, we modify the sphere of influence to depend upon the deterministic expected in-degree, instead of the stochastic actual in-degree. First we present a theorem from [10] on the expected in-degree of a vertex in the SPA model.

Theorem 1. *Let $\omega = \omega(t)$ be any function tending to infinity together with t. The expected in-degree at time t of a vertex v_i born at time $i \geq \omega$ is given by*

$$\mathbb{E}(\deg^-(v_i, t)) = (1 + o(1))\frac{A_2}{A_1}\left(\frac{t}{i}\right)^{A_1} - \frac{A_2}{A_1}$$

A vertex's birth time i and the size of the overall network t determines its expected in-degree. Furthermore, if the expected degree is larger than $\log^2 n$ then the real in-degree is concentrated around its expected in-degree [10]. Hence, the time a vertex is born can be used as a proxy for its degree, with old nodes being more highly connected than young nodes.

We modify the SPA model by redefining the sphere of influence to depend on the vertex's expected in-degree, instead of its actual in-degree. This substitution makes the size of the sphere of influence a nonrandom object. Precisely, the *modified spatial preferential attachment* model is defined as the SPA model, with the one difference being that the size of the sphere of influence of vertex v_i at time is changed to:

$$|S(v_i, t)| = \min\left\{\frac{A_2}{t^{1-A_1}i^{A_1}}, 1\right\} \tag{2}$$

We derive Eq. (2) by replacing the actual in-degree in formula (1) for the original sphere of influence with the expected in-degree and simplifying. We now state and prove a theorem which shows that the modified SPA model generates networks with an expected in-degree equivalent to the original model.

Theorem 2. *Let G_n be a graph generated by the modified SPA model with n vertices. The expected in-degree of a vertex v_i born at time i is given by*

$$\mathbb{E}(\deg^-(v_i, n)) = \frac{A_2}{A_1}\left(\left(\frac{n}{i}\right)^{A_1} - 1\right) - \epsilon$$

where $|\epsilon| < \frac{A_2}{A_1}$.

Proof. Let v_j be a vertex born at time $j > i$. Let X_j be a random variable that equals 1 if there is an edge from v_j to v_i and equals 0 otherwise. By the definition of the modified SPA model, we place an edge from v_j to v_i if and only if v_j falls within the sphere of influence of v_i. Since v_j is placed in the hypercube uniformly at random, $X_j = 1$ with probability equal to the volume of v_i's sphere of influence at time j.

By the linearity of expectation,

$$\mathbb{E}(\deg^-(v_i, n)) = \mathbb{E}\left(\sum_{k=i+1}^{n} X_k\right)$$

$$= \sum_{k=i+1}^{n} \mathbb{E}(X_k)$$

$$= \sum_{k=i+1}^{n} |S(v_i, k)|$$

We approximate this sum with an integral.

$$\sum_{k=i+1}^{n} |S(v_i, k)| = \int_{i}^{n} \frac{A_2}{x^{1-A_1} i^{A_1}} dx - \epsilon$$

$$= \frac{A_2}{A_1}\left(\left(\frac{n}{i}\right)^{A_1} - 1\right) - \epsilon$$

To bound the error, we first recognize that,

$$\int_{i+1}^{n+1} \frac{A_2}{x^{1-A_1} i^{A_1}} dx \leq \sum_{k=i+1}^{n} |S(v_i, k)| \leq \int_{i}^{n} \frac{A_2}{x^{1-A_1} i^{A_1}} dx$$

Hence,

$$|\epsilon| < \int_{i}^{n} \frac{A_2}{x^{1-A_1} i^{A_1}} dx - \int_{i+1}^{n+1} \frac{A_2}{x^{1-A_1} i^{A_1}} dx$$

$$= \frac{A_2}{i^{A_1}} \frac{((i+1)^{A_1} - i^{A_1}) - ((n+1)^{A_1} - n^{A_1})}{A_1}$$

$$< \frac{A_2}{i^{A_1}} \frac{(i+1)^{A_1} - i^{A_1}}{A_1}$$

$$< \frac{A_2}{A_1} (2^{A_1} - 1)$$

$$< \frac{A_2}{A_1}$$

$\square$

In addition to having equivalent expected in-degrees, we also derive that both models lead to the same (power law) cumulative in-degree distribution. The cumulative in-degree distribution c_k is defined as the number of vertices with in-degree $j \leq k$ divided by the total number of vertices. As stated above, from [1] we know that a.a.s. the cumulative in-degree distribution of networks generated by the SPA model follows a power law with exponent $\frac{1}{A_1}$. Theorem 3 states that the same is true of networks generated by the modified SPA model. An event occurs *with extreme probability (w.e.p.)* if it occurs with probability at least $1 - e^{-\Theta(\log^2 n)}$ as $n \to \infty$.

Theorem 3. *Let G_n be a graph generated by the modified SPA model with n vertices. The cumulative in-degree distribution c_k of G_n is w.e.p. a power law with exponent $\frac{1}{A_1}$ for $k > k' = \log^2(n)$.*

Proof. The in-degree of a vertex v_i born at time i is the sum of $n - i$ independent Bernoulli variables X_j with success probability equal to the volume of the sphere of influence $|S(v_i, j)|$. We let $f(i)$ equal the expected in-degree of a vertex born at time i in a network of size n. By the generalized Chernoff bound [11], we know that w.e.p. $f(i) - \epsilon \leq \deg^-(v_i, n) \leq f(i) + \epsilon$ where $\epsilon = \sqrt{f(i)} \log n$.

Using this bound, we determine how many vertices have in-degree greater than k. If a vertex is born before $i^- = f^{-1}(k + \epsilon)$, then w.e.p. it has a degree greater than k. Likewise, if a vertex is born after $i^+ = f^{-1}(k - \epsilon)$, then w.e.p. it has a degree less than k. Hence, the number of vertices with degree greater than k is between i^- and i^+. By inverting the formula for expected degree and examining its asymptotic growth, we find

$$i^- = f^{-1}(k + \epsilon) = (1 + o(1))f^{-1}(k) \qquad i^+ = f^{-1}(k - \epsilon) = (1 + o(1))f^{-1}(k)$$

Hence, the number of vertices with degree greater than k is w.e.p. $(1 + o(1))f^{-1}(k)$. This implies that $c_k = (1 + o(1))(kA_1/A_2 - 1)^{-1/A_1}$. $\qquad\square$

2.3 Infectious Processes

Now that we have a workable model of real-world networks, we define a SIR disease model in discrete time. To begin the infectious process, we pick a node at random to be the origin node. At time $t = 0$, we infect the origin node and denote all other nodes as susceptible. In each time step, the infected nodes infect each neighbour with probability β. Though the modified SPA model generates directed graphs, we ignore the orientation of the edges. If a susceptible vertex has multiple infected neighbours, they each independently attempt to infect the susceptible vertex. At the end of each time step, all infected nodes recover. This is a simplification of the typical SIR model because usually the recovery time is modelled as a stochastic variable. Here we simplify and assume each vertex to recover in exactly one time step. We run the process until no vertices are infected. If we run the infection process for t time steps on a network with n vertices, the infected and recovered vertices together with the edges taken by the infection (oriented in the direction the infection travelled) form an acyclic directed subgraph of the network. We denote this subgraph, I_n^t, the *infection graph at time t*.

Suppose that in a given time step vertex v is infected, vertex u is susceptible, and they are neighbours. The probability β of v infecting u in the time step depends on $\kappa(v)$, the average number of contacts v makes with u per time step, and the probability of transmitting the infection per contact τ. If v makes more contacts on average with u or if the probability of infection per contact is higher, the disease will have a greater chance of spreading. Following [5], we set

$$\beta = 1 - e^{-\tau\kappa}$$

To study how the behaviour of high degree nodes changes how a disease spreads through the network, we consider two different scenarios: scenario A and scenario B with their own infection probabilities β_A and β_B, respectively. In scenario A, we define the average number of contacts a vertex v makes with a neighbour per time step as

$$\kappa_A(v) = \frac{T}{\mathbb{E}(\deg^-(v))}$$

where T is the average number of contacts v makes with *all* its neighbours in the time step. We use $\mathbb{E}(\deg^-(v))$ as a rough approximation of the degree of v. Hence, in scenario A, $\beta_A(v) = 1 - e^{-\tau\kappa_A(v)}$. Since T is constant for all vertices, high and low degree alike, high degree vertices make fewer contacts with any *single* neighbour because their contacts are dispersed over a greater number vertices. While high degree nodes have many neighbours, these connections may be weaker than a node with few neighbours.

In scenario B, however, we no longer keep the average number of contacts a vertex makes with all its neighbours in a time step constant. Instead, we let $T(v)$ depend on T and the expected in-degree of v. We define

$$T(v) = T\frac{\mathbb{E}(\deg^-(v))}{\langle\deg^-\rangle}$$

where $\langle\deg^-\rangle$ is the average degree in the graph. From [1], we know a.a.s. that $\langle\deg^-\rangle = (1 + o(1))\frac{A_2}{1-A_1}$ in the SPA model, which is asymptotically constant. Since the total number of edges in the network is the sum of Bernoulli variables, by the linearity of expectation, it is a simple calculation to show that $\langle\deg^-\rangle$ is equivalent in the modified and original SPA models. As in scenario A, we use $\mathbb{E}(\deg^-(v))$ to approximate the degree of v. From the average number of contacts a vertex makes with *all* its neighbours in a time step, we can define the average number of contacts a vertex makes with a *specific* neighbour in a time step as

$$\kappa_B(v) = \frac{T(v)}{\mathbb{E}(\deg(v))} = \frac{T}{\langle\deg^-\rangle}$$

Hence, in scenario B, $\beta_B = 1 - e^{-\tau\kappa_B}$, which is constant. Any infected vertex v has an equal chance of infecting a neighbour, regardless of the degree of v. Hence, in scenario B, we expect that high degree vertices will pass the infection on to more individuals than low degree vertices.

3   Spatial Spread of Infections

Our main result states that when highly connected vertices are less infectious, the infection will not make large jumps through the metric space. Since our metric space represents a feature space, this means that the infection spreads through communities rather than jumping between them. To prove this result, we treat the infection as percolating through the network. We first show that a.a.s. all vertices in the network will only infect neighbours within a certain distance. From this result, we conclude that any particular infection will a.a.s. be bounded by a ball of a relatively small radius after a given number of time steps.

Theorem 4. *Let G_n be a graph with n vertices generated by the modified SPA model. Let $\lambda = n^{-\phi}$ be such that $\phi < \frac{A_1(1-A_1)}{(A_1+2)d}$. For scenario A, a.a.s. all nodes in the infection graph at time t will be within $t\lambda$ of the origin node u.*

3.1   Proof of Theorem 4

Before we present the proof, we first adopt some conventions regarding the infection process. Instead of considering the infection spreading through a network in time, we *a priori* consider whether any vertex would infect a neighbour given that they are connected. If we "occupy" each pair of vertices with probability β_A, and restrict our occupied pairs to edges present in our network, we get a subgraph consisting of where the infection could *possibly* travel, which we call the potential infection graph. The infection graph, describing where the infection *actually* spread, will be a subset of the potential infection graph.

Formally, let $G_n = (V_n, E_n)$ be a network of order n generated by the modified SPA model, where we replace each directed edge by two edges in opposite directions. We consider ordered pairs of vertices (v_i, v_j) and (v_j, v_i) because our infection model ignores the orientation of the edges in the original network generated by the modified SPA model. In other words, even though all edges in the modified SPA model are directed from younger to older vertices, we allow the old to infect the young. Let $u \in V_n$ be the node where the infection originates. With each ordered pair of vertices (v_i, v_j) we associate a Bernoulli random variable I_{v_i, v_j} defined as

$$I_{v_i, v_j} = \begin{cases} 1 & \text{with probability } \beta_A(v_i) \\ 0 & \text{otherwise} \end{cases} \tag{3}$$

We define the *potential infection graph* on G_n as the graph $I_n = (V_I, E_I)$ where $V_I = V_n$ and $E_I = \{(v_i, v_j) | \{v_i, v_j\} \in E_n \text{ and } I_{v_i, v_j} = 1\}$. We can recover the infection graph at time t, I_n^t, by taking the subgraph induced by the tth out-neighbourhood of u in I_n.

The proof of our main result is based on an analysis of the edges in the potential infection digraph. Define the *length of an edge* as the distance between its two end points. We first establish a lemma stating that there is an asymptotic bound on the length of edges in the potential infection graph.

Lemma 1. *Let G_n be a graph with n vertices generated by the modified SPA model and I_n be a potential infection graph on G_n in scenario A. Let $\lambda = n^{-\phi}$ such that $\phi < \frac{A_1(1-A_1)}{(A_1+2)d}$. Then a.a.s. I_n does not contain any edges of length greater than λ.*

Proof. Let L represent the event of there being an edge in I_n where the distance between its endpoints is greater than λ. We will call such edges "long" and all other edges "short". Given two (not necessarily connected) nodes in V_n, v_i and v_j, let L_{v_i,v_j} represent the event of there being a long edge from v_i to v_j in I_n. Thus, L_{v_i,v_j} occurs if v_i and v_j (the vertices born at time i and j, respectively) have distance at least λ, there is an edge between v_i and v_j, and the infection can travel from v_i to v_j. Since $L = \bigcup_{i=0}^{n-1} \bigcup_{j=i+1}^{n} \left(L_{v_i,v_j} \cup L_{v_j,v_i} \right)$, by taking the union bound, we know

$$\mathbb{P}(L) \leq \sum_{i=0}^{n-1} \sum_{j=i+1}^{n} \mathbb{P}(L_{v_i,v_j}) + \mathbb{P}(L_{v_j,v_i})$$

Our proof will show that this double sum goes to 0 as n approaches infinity.

We first need an expression for $\mathbb{P}(L_{v_i,v_j}) + \mathbb{P}(L_{v_j,v_i})$. Since $i < j$, by the definition of the potential infection graph, L_{v_i,v_j} occurs if and only if three other events also occur: $d(v_i, v_j) > \lambda$, $v_j \in S(v_i, j)$, and $I_{v_i,v_j} = 1$. In other words, for there to be a long edge between v_i and v_j, I_{v_i,v_j} must equal 1 and v_j must be far enough away from v_i to be considered long, but close enough to be in the sphere of influence of v_i at time j.

Since v_j is placed uniformly at random in the hypercube, the distance $d(v_j, v_i)$ and the event $I_{v_i,v_j} = 1$ are independent. Therefore, for any specific values for j and i. $i < j$, we can write

$$\mathbb{P}(L_{v_i,v_j}) = \mathbb{P}(d(v_i, v_j) > \lambda, v_j \in S(v_i, j), I_{v_i,v_j} = 1)$$
$$= \mathbb{P}(d(v_i, v_j) > \lambda, v_j \in S(v_i, j)) \mathbb{P}(I_{v_i,v_j} = 1)$$

For the edge oriented in the opposite direction, we can make a similar argument. Hence, we can write

$$\mathbb{P}(L_{v_j,v_i}) = \mathbb{P}(d(v_i, v_j) > \lambda, v_j \in S(v_i, j), I_{v_j,v_i} = 1)$$
$$= \mathbb{P}(d(v_i, v_j) > \lambda, v_j \in S(v_i, j)) \mathbb{P}(I_{v_j,v_i} = 1)$$

Combining our expressions for $\mathbb{P}(L_{v_i,v_j})$ and $\mathbb{P}(L_{v_j,v_i})$ we find that

$$\mathbb{P}(L_{v_i,v_j}) + \mathbb{P}(L_{v_j,v_i}) = \mathbb{P}(d(v_i, v_j) > \lambda, v_j \in S(v_i, j))(\mathbb{P}(I_{v_i,v_j} = 1) + \mathbb{P}(I_{v_j,v_i} = 1))$$

We know $\mathbb{P}(I_{v_i,v_j} = 1) = \beta_A(v_i)$ and $\mathbb{P}(I_{v_j,v_i} = 1) = \beta_A(v_j)$ from Eq. (3), but we need expressions for $\mathbb{P}(d(v_i, v_j) > \lambda, v_j \in S(v_i, j))$ and $\mathbb{P}(I_{v_i,v_j} = 1)$.

We use a geometric argument to find $\mathbb{P}(d(v_j, v_i) > \lambda, v_j \in S(v_i, j))$, which is the probability of there being a long edge between v_i and v_j in the original network. There are three cases. In the first case, the sphere of influence of v_i has radius smaller than λ at its time of birth. Since spheres of influence only shrink, in this case there will never be a time when v_j can both fall within v_i's sphere of influence and be greater than λ away from v_i. This case occurs when i exceeds a critical value m, which is the first time when vertices are born with a sphere of influence that has radius smaller than λ.

In the second case, i is smaller than the critical value m, but j is larger than the second critical value m_i. This critical value is reached when the radius of v_i's sphere of influence equals λ. Again, since spheres of influence only shrink, vertices born at times after m_i cannot have $d(v_j, v_i) > \lambda$ and $v_j \in S(v_i, j))$. In these first two cases, $\mathbb{P}(d(v_j, v_i) > \lambda, v_j \in S(v_i, j)) = 0$.

A ball of radius λ has volume $\lambda^d c_p$ where c_p depends on our L_p norm. Using this, we find that

$$m = \frac{A_2}{\lambda^d c_p} \qquad m_i = \left(\frac{A_2}{i \lambda^d c_p} \right)^{\frac{1}{1-A_1}}$$

In the last case, $i < m$ and $j < m_i$, which means $d(v_j, v_i) > \lambda$ and $v_j \in S(v_i, j))$ is possible. Since v_j is placed in the hypercube uniformly at random and the hypercube has unit volume, $\mathbb{P}(d(v_j, v_i) > \lambda, v_j \in S(v_i, j))$ is equal to the volume of the spherical shell between the sphere of influence and the ball centered at v_i with radius λ which we denote $B(v_i, \lambda)$. Hence, in this case, $\mathbb{P}(d(v_j, v_i) > \lambda, v_j \in S(v_i, j)) = |S(v_i, j)| - |B(v_i, \lambda)|$.

Combining the results from the previous paragraphs,

$$\mathbb{P}(L) \leq \sum_{i=0}^{m} \sum_{j=i+1}^{m_i} (|S(v_i, j)| - |B(v_i, \lambda)|)(\beta_A(v_i) + \beta_A(v_j))$$

Since the oldest vertex will always have the largest sphere of influence, we know that $m_i < m_1$ for all $i \in [1, m]$ and that $m < m_1$. Also, $A_2 = |S(v_1, 1)| > |S(v_i, j) - B(v_i, \lambda)|$ for all $i, j \in [1, m_1]$. Finally, since v_{m_1} has the lowest expected degree of all vertices born at or before time m_1, $\beta_A(v_{m_1}) > \beta_A(v_i), \beta_A(v_j)$ for all $i, j \in [1, m_1]$. Hence, we can write

$$\mathbb{P}(L) \leq \sum_{i=1}^{m_1} \sum_{j=i+1}^{m_1} 2A_2 \beta_A(v_{m_1}) \tag{4}$$

$$\leq 2 \left(1 - \exp \left(-\frac{\tau T}{\frac{A_2}{A_1} \left(\left(\frac{n}{m_1} \right)^{A_1} - 1 \right)} \right) \right) A_2 m_1^2 \tag{5}$$

From the formula for m_1, we see that $m_1 \sim n^{\frac{\phi d}{1-A_1}}$. Setting $\phi = \frac{A_1(1-A_1)}{(A_1+2)d}$ and $\gamma = \tau T$, we see that

$$\mathbb{P}(L) \leq 2 \left(1 - \exp\left(-\frac{\gamma}{\left(\left(\frac{n}{m_1}\right)^{A_1} - 1 \right)} \right) \right) A_2 m_1^2 \tag{6}$$

$$\sim 2 \left(1 - \exp\left(-\gamma n^{A_1\left(\frac{\phi d}{1-A_1}-1\right)} \right) \right) A_2 n^{\frac{2\phi d}{1-A_1}} \tag{7}$$

$$\sim 2 \left(1 - \left(1 - \gamma n^{A_1\left(\frac{\phi d}{1-A_1}-1\right)} + O\left(n^{2A_1\left(\frac{\phi d}{1-A_1}-1\right)} \right) \right) \right) A_2 n^{\frac{2\phi d}{1-A_1}} \tag{8}$$

$$= o(1) \tag{9}$$

$\square$

Using this lemma, we can now prove our main result, Theorem 4.

Proof. Let B represent the bad event where there is a node v in the infection digraph at time t, I_n^t, where $d(v,u) > t\lambda$. If B occurs, then there is a path from v to u with at most t edges because I_n^t is the tth neighbourhood of u in I_n. By the triangle inequality, at least one of the edges in the path from v to u has a length greater than λ and, more generally, there is an edge in the potential infection graph with a length greater than λ. Let L represent the event of there being an edge in I_n where the distance between its endpoints is greater than λ. Since $B \subset L$, $\mathbb{P}(B) \leq \mathbb{P}(L)$, but by the previous lemma, a.a.s. $\mathbb{P}(L) = 0$. $\square$

3.2 Conjecture for Scenario B

We conjecture that in scenario B, the negation of Lemma 1 holds. We know that the modified SPA model a.a.s. has edges greater than length λ' where $\lambda' = \mu n^{-\theta}$ with $\theta > 1 - \frac{A_1}{4A_1+2}$ and μ constant. We conjecture that the potential infection graph will have long edges as well.

Conjecture 1. *Let G_n be a graph with n vertices generated by the modified SPA model and I_n be a potential infection graph on G_n in scenario B. There exists $\phi > 0$ such that if we let $\lambda = n^{-\phi}$, a.a.s. I_n contains an edge of length greater than λ.*

4 Simulations

Using simulations, we test our theoretic result that the infection in scenario A will not make long jumps. We also use simulations to provide evidence for our conjecture that, in scenario B, the infection can make long jumps if we pick the origin vertex correctly. Recall that in both infection scenarios, we can vary how easily the infection spreads by altering T, the total number of contacts made per time step, and τ, the probability of transmission per contact. Also recall that we denote $\gamma = \tau T$. For the 2 infection scenarios, we consider 3 levels of contagiousness: $\gamma = 1$, $\gamma = 10$ and $\gamma = 100$.

We generated 10 networks with the modified SPA network in $\mathbb{R}^1$ with $A_1 = 0.5$ and $A_2 = 1$. Our results are highly asymptotic and the bound is lowest in low dimensions so, due to computational constraints, we choose to simulate in $\mathbb{R}^1$. The 10 networks are of increasing size, beginning at $n = 1000$ and increasing by increments of 1000 to a maximum of $n = 10000$.

For each network, we run each of the 6 infection processes 50 times. We chose to begin the infections at the oldest vertex because it has the highest likelihood of having neighbours far away in the metric space. On one hand, we want to give the infection in scenario B the opportunity to make long jumps and, on the other hand, we do not want to mistakenly conclude that the infection in scenario A makes short jumps only because it was never exposed to long edges. While our main result states that given a number of time steps, a.a.s. the infection remains within a certain region, this result depends on both the size of the network and the current time step in the infection process. We thought it would be more clear to compare Lemma 1 to our simulations, which states that the edge length taken by the infection in scenario A is bounded. To compare scenario A and B, we likewise observe the maximum edge length the infection in scenario B takes.

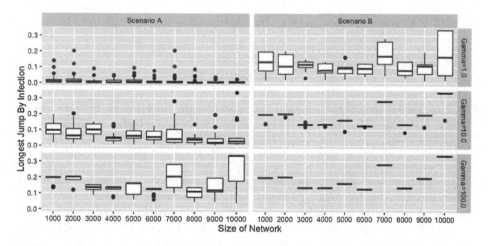

Fig. 1. Longest jump made by the infection vs. network size in scenario A and B stratified by 3 levels of contagion.

The results of our simulations are shown in Fig. 1. We make two conclusions from our simulations. First, we see that in scenario A, when high degree nodes are less contagious, the infection takes shorter jumps than in scenario B. As indicated by our asymptotic result, the difference becomes more pronounced in larger networks. One might notice that in scenario B the infection does not always make long jumps, which seems to contradict our conclusion. These outliers can be explained by recognizing that the longest edge in the entire network is a non-clustered random variable. Vertices receive long edges in a brief period during the early steps of the model and, consequently, whether there are long edges in the network at all is highly variable. It is not that the infection avoids the long edges, but rather, that the infection has no long edges to take in the first place. Second, we see that this difference between scenario A and scenario B becomes less pronounced when we make the infection more contagious. Again, this matches our theoretic result, since our bound on the probability of long infection increases when γ is larger. We expect that if we could generated large enough networks, eventually we would see the difference between scenario A and B reemerge, even at high levels of contagion.

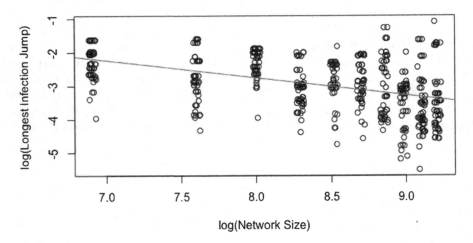

Fig. 2. A log-log plot of the longest jump made by the infection vs. network size in scenario A with $\gamma = 10$. The equation for the regression line is $\log(y) = -0.51\log(x) + 1.40$ ($R^2 = 0.19$).

To compare our analytic bound ϕ to our simulations, we perform a linear regression on a log-log plot of longest jump vs. network size for infections in Scenario A with $\gamma = 10$ (see Fig. 2). The simulations are the same as those represented in Fig. 1. We have added a small amount of noise to the x-values in order to make the distribution of data more clear. If our result is true, then we should expect that the slope of the regression line should be less than $-\frac{A_1(1-A_1)}{(A_1+2)d} = -0.1$. We found the slope to be -0.51, which provides support for our lemma. Of course, our data is highly variable and this plot only gestures towards the fact that our bound is valid. We expect that for larger networks, this variation would decrease.

To illustrate that the infection spreads slowly through the feature space in scenario A, we simulate one run of the process on a graph generated by the modified SPA model in $\mathbb{R}^2$ with $A_1 = 0.5$, $A_2 = 1$, $\gamma = 10$, and $n = 1000$. We present the simulation in Fig. 3. The blue vertices were infected earlier in

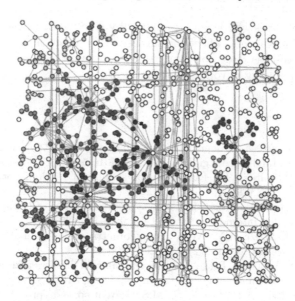

Fig. 3. A scenario A infection with $\gamma = 10$ on a modified SPA network with $A_1 = 0.5$, $A_2 = 1$, and $n = 1000$. The gradient from blue to red represents earlier to later infections. (Color figure online)

the process and red vertices later. Since nearby vertices have similar colours (recall that we are using the torus metric), this simulation provides additional qualitative evidence for our finding that the infection does not make long jumps in scenario A.

5   Conclusion

When modelling contagious processes, it is important to take contacts made per neighbour into account. With analytic and numeric results, we show that if all vertices make an equal number of contacts, the infection will spread through communities rather than jumping between them. High degree vertices are more likely to have neighbours in distant communities and when these popular individuals are less contagious, the infection is less likely to spread from one community to another. We also show with simulations that scaling the number of contacts a vertex makes by its degree results in an epidemic that spreads irrespective of the communities in the network.

In addition to our conjecture, we identify two areas of future research. First, since infections in scenario A and scenario B behave differently with respect to community structure, interventions may benefit from exploiting this distinction. In other words, if we know a contagious process will spread through communities, how can we use this fact to control the epidemic? Likewise, how should we control diseases that jump between communities? The second area of potential research is studying scenario A infections further. While we find that these types of contagious processes will be a.a.s. bounded by a ball with a growing radius, we do not discuss how this may affect the success of an infection spreading through a network. If a disease does not jump, does community structure prevent the infection from spreading beyond its original group?

References

1. Aiello, W., Bonato, A., Cooper, C., Janssen, J., Pralat, P.: A spatial web graph model with local influence regions. Internet Math. **5**(12), 173–193 (2007)
2. Barabási, A.L., Albert, R.: Emergence of scaling in random networks. Science **286**(5439), 509–512 (1999)
3. Cooper, C., Frieze, A., Pralat, P.: Some typical properties of the spatial preferred attachment model. Internet Math. **10**(1–2), 116–136 (2014)
4. Flaxman, A.D., Frieze, A.M., Vera, J.: A geometric preferential attachment model of networks. Internet Math. **3**(2), 187–206 (2006)
5. Garnett, G.P., Anderson, R.M.: Sexually transmitted diseases and sexual behavior: insights from mathematical models. J. Infect. Dis. **174**(2), S150–S161 (1996)
6. Hoff, P.D., Raftery, A.E., Handcock, M.S.: Latent space approaches to social network analysis. J. Am. Stat. Assoc. **97**(460), 1090–1098 (2002)
7. Jacob, E., Morters, P.: Spatial preferential attachment networks: power laws and clustering coefficients. Ann. Appl. Probab. **25**(2), 632–662 (2015)
8. Janssen, J., Hurshman, M., Kalyaniwalla, N.: Model selection for social networks using graphlets. Internet Math. **8**(4), 338–363 (2012)
9. Janssen, J., Mehrabian, A.: Rumours spread slowly in a small world spatial network. In: Gleich, D.F., Komjáthy, J., Litvak, N. (eds.) WAW 2015. LNCS, vol. 9479, pp. 107–118. Springer, Cham (2015). doi:10.1007/978-3-319-26784-5_9
10. Janssen, J., Pralat, P., Wilson, R.: Geometric graph properties of the spatial preferred attachment model. Adv. Appl. Math. **50**(2), 243–267 (2013)
11. Lu, L., Chung, F.: Old and new concentration inequalities. Complex Graphs and Networks, Chap. 2, pp. 23–56. American Mathematical Society, Providence (2006)
12. Newman, M.E.J., Peixoto, T.P.: Generalized communities in networks. Phys. Rev. Lett. **115**(8), 08871 (2015)
13. Nordvik, M.K., Liljeros, F.: Number of sexual encounters involving intercourse and the transmission of sexually transmitted infections. Sex. Transm. Dis. **33**(6), 342–349 (2006)
14. Salathé, M., Jones, J.H., May, R., Johnson, A., Auranen, K.: Dynamics and control of diseases in networks with community structure. PLoS Comput. Biol. **6**(4), e1000736 (2010)

Preferential Placement for Community Structure Formation

Aleksandr Dorodnykh[1], Liudmila Ostroumova Prokhorenkova[1,2(✉)],
and Egor Samosvat[1,2]

[1] Moscow Institute of Physics and Technology, Moscow, Russia
[2] Yandex, Moscow, Russia
Ostroumova-la@yandex.ru

Abstract. Various models have been recently proposed to reflect and
predict different properties of complex networks. However, the community structure, which is one of the most important properties, is not
well studied and modeled. In this paper, we suggest a principle called
"preferential placement", which allows to model a realistic community
structure. We provide an extensive empirical analysis of the obtained
structure as well as some theoretical heuristics.

1 Introduction

The evolution of complex networks attracted a lot of attention in recent years.
Empirical studies of different real-world networks have shown that such structures have some typical properties: small diameter, power-law degree distribution, clustering structure, and others [9,15,35]. Therefore, numerous random
graph models have been proposed to reflect and predict such quantitative and
topological aspects of growing real-world networks [9,11,15,38,41].

The most extensively studied property of complex networks is their vertex
degree distribution. For the majority of studied real-world networks, the portion
of vertices of degree d was observed to decrease as $d^{-\gamma}$, usually with $2 < \gamma < 3$ [5,
18,36]. Such networks are often called scale-free. The most well-known approach
to the modeling of scale-free networks is called *preferential attachment*. The main
idea of this approach is that new vertices emerging in a graph connect to some
already existing vertices chosen with probabilities proportional to their degrees.
Preferential attachment is a natural process allowing to obtain a graph with a
power-law degree distribution, and many random graph models are based on
this idea, see, e.g., [10,13,23,26,45].

Another important characteristic of complex networks is their community (or
clustering) structure, i.e., the presence of densely interconnected sets of vertices,
which are usually called clusters or communities [19,21]. Several empirical studies
have shown that community structure of different real-world networks has some
typical properties. In particular, it was observed that the cumulative community
size distribution obeys a power law with some parameter λ. For instance, [14]
reports that $\lambda = 1$ for some networks; [3] obtains either $\lambda = 0.5$ or $\lambda = 1$; [22]

© Springer International Publishing AG 2017
A. Bonato et al. (Eds.): WAW 2017, LNCS 10519, pp. 75–89, 2017.
DOI: 10.1007/978-3-319-67810-8_6

also observes a power law with λ close to 0.5 in some range of cluster sizes; [39] studies the overlapping communities and shows that λ is ranging between 1 and 1.6.

Community structure is an essential property of complex networks. For example, it highly affects the spreading of infectious diseases in social networks [24,29], spread of viruses over computer networks [43], promotion of products via viral marketing [25], propagation of information [42], etc. Therefore, it is crucial to be able to model realistic community structures.

Nowadays, there are a few random graph models allowing to obtain clustering structures. Probably the most well-known model was suggested in [28] as a benchmark for comparing community detection algorithms. In this model, the distributions of both degrees and community sizes follow power laws with predetermined exponents. However, there are two drawbacks of this model. First, it does not explain the power-law distribution of community sizes, these sizes are just sampled from a power-law distribution at the beginning of the process. Second, a subgraph induced by each community is very similar to the configuration model [8], which does not allow to model, e.g., hierarchical community structure often observed in real-world networks [3,14].

A weighted model which naturally generates communities was proposed in [27]. However, the community structure in this model is not analyzed in details and only the local clustering coefficient is shown. From the figures presented in [27] it seems that the community size distribution does not have a heavy tail as it is observed in real-world complex networks.

Finally, let us mention a paper [40] which analyzes a community graph, where vertices refer to communities and edges correspond to shared members between the communities. The authors show that the development of the community graph seems to be driven by preferential attachment. They also introduce a model for the dynamics of overlapping communities. Note that [40] only models the membership of vertices and does not model the underlying network.

In this paper, we propose a process which naturally generates clustering structure. Our approach is called *preferential placement* and it is based on the idea that vertices can be embedded in a multidimensional space of latent features. The vertices appear one by one and their positions are defined according to preferential placement: new vertices are more likely to fall into already dense regions. We present a detailed description of this process in Sect. 2. After n steps we obtain a set of n vertices placed in a multidimensional space. In Sect. 3 we empirically analyze the obtained structure: in particular, we show that the communities are clearly visible and their sizes are distributed according to a power law. Note that after the placement of all vertices is defined, one can easily construct an underlying network, using, e.g., the threshold model [12,31]. We discuss possible models and their properties in Sect. 4.

2   Preferential Placement

In this section, we describe the proposed approach which we call *preferential placement*. We assume that all vertices are embedded in $\mathbb{R}^d$ for some $d \geq 1$.

One can think that coordinates of this space correspond to latent features of vertices. Introducing latent features has recently become a popular approach both in predictive and generative models. These models are known by different names such as latent feature models [33,34], matrix factorization models [4,16, 32], spatial models [2,6,7], or geographical models [12,31]. The basic idea behind all these models is that vertices having similar latent features are more likely to be connected by an edge.

Preferential placement is the procedure describing the embedding of vertices in the space $\mathbb{R}^d$. After that, given the coordinates of all vertices, one can construct a graph using one of many well-known approaches (see Sect. 4 for the discussion of possible variants).

Our model is parametrized by a distribution Ξ taking nonnegative values. The proper choice of Ξ is discussed further in this section.

We construct a random configuration of vertices (or points) $S_n = \{\mathbf{x}_1, \ldots, \mathbf{x}_n\}$, where $\mathbf{x}_i = (x_i^1, \ldots, x_i^d)$ denotes the coordinates of the i-th vertex v_i. Let $S_1 = \{\mathbf{x}_1\}$, $\mathbf{x}_1$ is the origin. Now assume that we have constructed S_t for $t \geq 1$, then we obtain S_{t+1} by adding a vertex v_{t+1} with the coordinates $\mathbf{x}_{t+1}$ chosen in the following way:

- Choose a vertex $v_{i_{t+1}}$ from $v_1, \ldots, v_t$ uniformly at random.
- Sample ξ_{t+1} from the distribution Ξ.
- Sample a direction $\mathbf{e}_{t+1}$ from a uniform distribution on a multidimensional sphere $\|\mathbf{e}_{t+1}\|_2 = 1$, where $\| \cdot \|_2$ denotes the Euclidean distance in $\mathbb{R}^d$.
- Set $\mathbf{x}_{t+1} = \mathbf{x}_{i_{t+1}} + \xi_{t+1} \cdot \mathbf{e}_{t+1}$.

We argue in this paper that in order to obtain a realistic clustering structure one should take Ξ to be a heavy tailed distribution. In this case, according to the procedure described above, new vertices will usually appear in the dense regions, close to some previously added vertices; however, due to the heavy tail of Ξ, from time to time we get outliers, which originate new clusters.

We call the described above procedure "preferential placement" due to its analogy with preferential attachment. Assume that at some step of the algorithm we have several clusters, i.e., groups of vertices located close to each other, and a new vertex appears. Then the probability that this vertex will join a cluster C is roughly proportional to its size, i.e., the number of vertices already belonging to this cluster. This is the basic intuition which we discuss further in this paper in more details.

3   Analysis of Preferential Placement

3.1   Experimental Setup

In this section, we analyze graphs obtained using the preferential placement procedure described above. We take Ξ to be a slightly modified Pareto distribution with the density function $f_\beta(x) = \frac{\beta}{(x+1)^{\beta+1}}, x \geq 0$ for fixed $\beta > 0$.

In all the experiments we take $d = 2$ since the obtained structures are easy to visualize. However, we also tried other values of $d \geq 1$ and obtained results

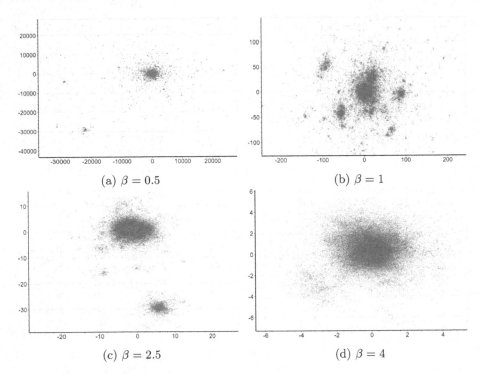

(a) $\beta = 0.5$

(b) $\beta = 1$

(c) $\beta = 2.5$

(d) $\beta = 4$

Fig. 1. Clustering structure depending on $\varXi$

similar to shown on Figs. 4 and 6. Also, if not specified otherwise, we generated structures with the number of points $n = 100\,\mathrm{K}$.

3.2   Clustering Structure Depending on $\varXi$

First, let us visualize the structures obtained by our algorithm. We tried several values of β, $\beta \in \{0.5, 1, 1.5, 2.5, 4\}$. The results are presented on Figs. 1 and 2. The value $\beta = 0.5$ produces the heaviest tail, in this case the distribution $\varXi$ does not have a finite expectation. Although some clusters are clearly visible in this case, they are located far apart from each other, which seems to be not very realistic. Graphs obtained from configurations (using one of the procedures discussed in Sect. 4) are expected to have small diameter and giant connected component of size $\Theta(n)$, which does not seem to be the case for $\beta = 0.5$. Note that for too large β, e.g., for $\beta = 4$, the variance is too low and we obtain only one giant cluster with minor fluctuations, as presented on Fig. 1d. Further in this paper we discuss the case $\beta = 1.5$ presented on Fig. 2. In this case $\varXi$ has a finite expectation but an infinite variance.

Another interesting observation is a hierarchical clustering structure produced by our algorithm. To illustrate this, we take the figure obtained for $\beta = 1.5$ and zoom it to see more details. Figure 2 shows that the largest cluster further consists of several sub-clusters.

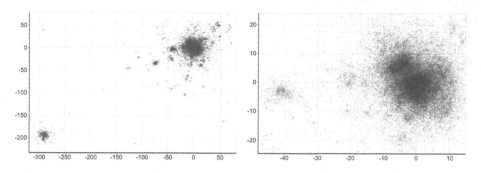

Fig. 2. $\beta = 1.5$, different scales

3.3 The Distribution of Cluster Sizes

In this section, we analyze the distribution of cluster sizes produced by preferential placement. We present both theoretical and empirical observations.

The main difficulty with the analysis of clustering structure is the fact that there are no standard definitions of clusters, both in graphs and metric spaces. For example, clusters are often defined as a result of some clustering algorithm.[1] This causes a lot of difficulties for both theoretical and empirical analysis.

Theoretical Heuristics. First, let us discuss why we expect to observe a power-law distribution of cluster sizes in our model. As we discussed above, due to the absence of a rigorous definition of a cluster, further in this section we are able to present only some heuristic theory.

Let $F_t(s)$ denote the number of clusters of size s at step t. In order to analyze $F_t(s)$ we consider its dynamics inductively. Assume that after a step t we obtain some clustering structure. At step $t + 1$ we add a vertex v_{t+1} and choose its "parent" $v_{i_{t+1}}$ from $v_1, \ldots, v_t$ uniformly at random. Clearly, the probability to choose a parent from some cluster C with $|C| = s$ is equal to $\frac{s}{t}$. In this case, we call C a parent cluster for v_{t+1}. Now let us make the following assumptions:

1. All clusters can only grow, they cannot merge or split.
2. At step $t + 1$ a new cluster appears with probability $p(t) = \frac{c}{t^\alpha}$, $c > 0$, $0 \le \alpha \le 1$.
3. Given that a vertex $t + 1$ does not create a new cluster, the probability to join a cluster C with $|C| = s$ is equal to $\frac{s}{t}$.

These assumptions are quite strong and even not very realistic. For instance, it seems reasonable that two clusters can merge if many vertices appear somewhere between them. Regarding the second assumption, $p(t)$ can possibly depend

---

[1] *Modularity*, introduced in [37], can be used to define communities in graphs. However, this characteristic has certain drawbacks, as discussed in [20]. Moreover, modularity favors partitions with approximately equal communities, which contradicts the main idea of power-law distribution of community sizes.

on the current configuration S_t. However, these assumptions allow us to analyze the behavior of $F_t(s)$ formally. Namely, we prove the following theorem.

Theorem 1. *Under the assumptions described above the following holds.*

1. *If $\alpha = 0$, then*

$$\mathrm{E}F_n(s) = \frac{c\,(s-1)!\,\Gamma\left(2 + \frac{1}{1-c}\right)}{(2-c)\Gamma\left(s+1 + \frac{1}{1-c}\right)}\left(n + O\left(s^{\frac{1}{1-c}}\right)\right)$$

$$\sim \frac{c\,\Gamma\left(2 + \frac{1}{1-c}\right)}{(2-c)} \cdot \frac{n}{s^{1 + \frac{1}{1-c}}}.$$

2. *If $0 < \alpha \le 1$, then for any $\epsilon > 0$*

$$\mathrm{E}F_n(s) = \frac{c\,(s-1)!\,\Gamma(3-\alpha)}{(2-\alpha)\Gamma(s+2-\alpha)}\left(n^{1-\alpha} + O\left(n^{\max\{0,1-2\alpha\}}s^{1-\alpha+\epsilon}\right)\right)$$

$$\sim \frac{c\,\Gamma(3-\alpha)}{2-\alpha} \cdot \frac{n^{1-\alpha}}{s^{2-\alpha}}.$$

To sum up, if the probability $p(n)$ of creating a new cluster is of order $\frac{1}{n^\alpha}$ for $\alpha > 0$, then the distribution of cluster sizes follows a power law with parameter $2 - \alpha$ growing with $p(n)$ from 1 to 2; if $p(n) = c$, $0 < c < 1$, then the parameter grows with c from 2 to infinity. Recall that the parameter of the cumulative distribution is one less than discussed above. The proof of Theorem 1 is technical and we place it to Appendix.

Let us also explain why we do not consider $p(n)$ decreasing faster than $\frac{c}{n}$. It is natural to assume that a new cluster appears if a new vertex chooses a parent node near the border of some cluster and then ξ_{t+1} and $\mathbf{e}_{t+1}$ are chosen such that $\mathbf{x}_{t+1} = \mathbf{x}_{i_{t+1}} + \xi_{t+1} \cdot \mathbf{e}_{t+1}$ falls quite away from the parent cluster. This probability is roughly proportional to the number of vertices located near the borders of the clusters. Extreme case, 1 vertex, provides the bound $\frac{c}{n}$.

Finally, let us mention that in practice the probability $p(n)$ of creating a new cluster can depend not only on $\varXi$, but also on the definition of clusters. Further in this section we demonstrate that parameters of a clustering algorithm can affect the parameter of the obtained power law.

Empirical Analysis. As we already mentioned, there is no standard definition of a clustering structure. In many cases, clusters and communities are defined just as a result of some clustering algorithm. Therefore, we first analyze the performance of several clustering algorithms, then choose the most appropriate one and analyze clusters it produces.

We compare the following algorithms: k-means [30], EM (expectation maximization), and DBSCAN (density-based spatial clustering of applications with noise) [17]. For k-means and EM one has to specify the number of clusters.

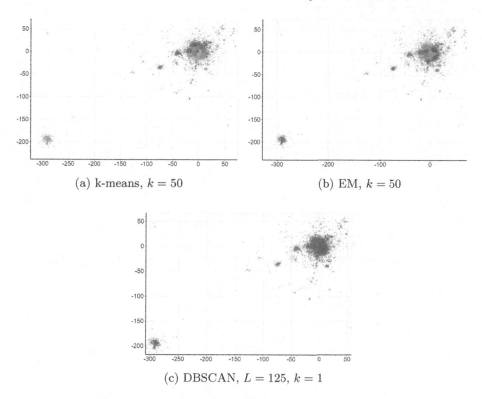

(a) k-means, $k = 50$ (b) EM, $k = 50$

(c) DBSCAN, $L = 125$, $k = 1$

Fig. 3. The comparison of different clustering algorithms

We tried several values of k, $k \in \{10, 50, 100, 500, 1000\}$, but both algorithms turned out to be not suitable for out problem. As expected, in all cases they unnaturally split the largest cluster into several small ones (see Figs. 3a and b).

On the contrary, DBSCAN produces more realistic results. It requires two parameters: radius of neighborhood ε and the minimum number of neighbors required to form a dense region k. We consider $k \in \{1, 2, 3\}$ and ε is chosen in such a way that if we connect all vertices i, j such that $\|i - j\|_2 < \varepsilon$, then we get Ln edges, $L \in \{5, 25, 125\}$, where n is the number of vertices. For all parameters we get reasonable clustering structures. The result for $L = 125, k = 1$ is presented on Fig. 3c. For these parameters we also analyze the distribution of cluster sizes (see Fig. 4a). Note that for not too large values of s ($s < 300$) the cumulative distribution follows a power law with parameter $\lambda \approx 0.95$. In Theorem 1 this value corresponds to the case $\alpha = 0.05$, i.e., $p(n) \propto n^{-0.05}$. Based on this, we expect the number of clusters to grow as $n^{0.95}$, i.e., close to linearly. On Fig. 5 we plot the empirical number of clusters and fit it by $n^{0.95}$.

Finally, as we promised above, we show that λ can depend on the clustering algorithm. Figure 4b shows the cumulative cluster size distribution for DBSCAN with $L = 5, k = 1$. Note that $\lambda = 1.44$, so it is larger in this case. Intuitively, the reason is that $p(n)$ is larger for $L = 5$ than for $L = 125$. Smaller values of L

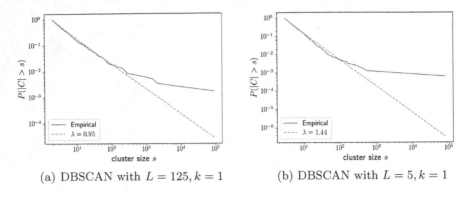

(a) DBSCAN with $L = 125, k = 1$ (b) DBSCAN with $L = 5, k = 1$

Fig. 4. Cluster size distribution

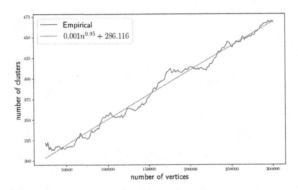

Fig. 5. Growth of the number of clusters, DBSCAN with $L = 125, k = 1$

correspond to smaller ε, which means that it is harder for a new vertex to join some existing cluster, which makes $p(n)$ larger.

4 Graph Models

4.1 Possible Definitions

In this section, we discuss how a graph can be constructed based on the vertices embedding produced by the preferential placement procedure.

The basic idea behind many known spatial models is that we want to increase the probability of connecting two vertices if they have similar latent features. Various methods can be found in the literature, which are usually combined with some other ideas like introducing weights of vertices or taking into account degrees of vertices (see, e.g., a survey of spatial models in [7]). We now briefly describe some possible approaches:

– *threshold model* [12,31]:

$$\mathrm{P}\big((v_i, v_j) \in E\big) = I\big[\|\mathbf{x}_i - \mathbf{x}_j\|_2 \leq \theta\big] ;$$

– *p-threshold model*:

$$P\big((v_i, v_j) \in E\big) = pI\big[\|\mathbf{x}_i - \mathbf{x}_j\|_2 \le \theta\big], \quad 0 < p < 1;$$

– *p-threshold model with random edges* (as in spatial small-world models [7]):

$$P\big((v_i, v_j) \in E\big) = p_0 + p_1 I\big[\|\mathbf{x}_i - \mathbf{x}_j\|_2 \le \theta\big], \quad 0 < p_0, p_1, p_0 + p_1 < 1;$$

– *inverted distance model*:

$$P\big((v_i, v_j) \in E\big) \propto \frac{1}{\|\mathbf{x}_i - \mathbf{x}_j\|_2};$$

– *Waxman model* [44]:

$$P\big((v_i, v_j) \in E\big) \propto e^{-\|\mathbf{x}_i - \mathbf{x}_j\|_2/d}.$$

Here we denote by E the set of edges. We assume that all edges are mutually independent, hence to describe a random graph it is enough to define the probability of each edge. Further we focus on the threshold model, however, we expect similar results for other models.

4.2 Degree Distribution

In this section, we empirically analyze the degree distribution for the threshold model. As before, we take Ξ to be a distribution with the density function $f_\beta(x) = \frac{\beta}{(x+1)^{\beta+1}}, x \ge 0$ for $\beta = 1.5$. We choose θ such that we have $5n$ edges in our graph. The cumulative degree distribution for this case is presented on Fig. 6. Observe that the cumulative degree distribution does not follow a power law. However, it is very similar to degree distributions obtained in many real-world networks (numerous examples can be found in [1]).

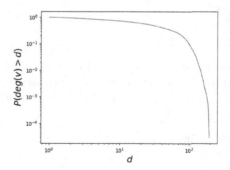

Fig. 6. Cumulative degree distribution for the threshold model

We are currently working on theoretical analysis of the degree distribution in the threshold model. We plan to add these results, together with the empirical analysis of other properties like diameter and clustering coefficient, to the extended version of this paper.

5   Conclusion and Future Work

In this paper, we introduced a principle called *preferential placement*. Our method is designed to model a realistic clustering structure. The algorithm is parametrized only by a distribution Ξ, and if Ξ is a Pareto distribution, which is the most natural choice, then we essentially have only one parameter — the exponent β. The proposed algorithm naturally models clusters and the distribution of cluster sizes follows a power law, which is a desirable property. Although preferential placement only generates the coordinates of vertices, one can easily construct a graph based on the obtained structure using one of the methods discussed in this paper. We showed that applying a threshold model to the configuration generated by preferential placement leads to a realistic degree distribution.

In this paper, we made only a first step to understanding the cluster formation in complex structures and there are many directions for future research. First of all, more formal analysis of the distribution of cluster sizes would be useful. As we discussed, the main problem here is the lack of any suitable formal definition of clusters. However, one can try, e.g., to analyze clusters produced by one of well-known clustering algorithms. Second direction is the analysis of the obtained graphs. We are currently working on theoretical analysis of the degree distribution in the threshold model. We also plan to analyze other properties, like diameter and clustering coefficient.

Acknowledgements. This work is supported by Russian President grant MK-527.2017.1.

Appendix

Proof of Theorem 1

First, recall the process of cluster formation:

- At the beginning of the process we have one vertex which forms one cluster.
- At n-th step with probability $p(n)$ a new cluster consisting of v_n is created.
- With probability $1 - p(n)$ new vertex joins already existing cluster C with probability proportional to $|C|$.

So, we can write the following equations:

$$\mathrm{E}(F_{t+1}(1)|S_t) = F_t(1)\left(1 - \frac{1 - p(t)}{t}\right) + p(t), \tag{1}$$

$$\mathrm{E}(F_{t+1}(s)|S_t) = F_t(s)\left(1 - \frac{s(1 - p(t))}{t}\right) + F_t(s-1)\frac{(s-1)(1 - p(t))}{t}, \quad s > 1. \tag{2}$$

Now we can take expectations of the both sides of the above equations and analyze the behavior of $\mathrm{E}F_t(s)$ inductively.

Consider the case $\alpha = 0$, i.e., $p(n) = c$. Let us prove that in this case

$$\mathrm{E}F_n(s) = \frac{c(s-1)!\,\Gamma\left(2 + \frac{1}{1-c}\right)}{(2-c)\Gamma\left(s+1+\frac{1}{1-c}\right)}\,(n + \theta_{n,s})\;. \qquad (3)$$

where $\theta_{n,s} \le C\,s^{\frac{1}{1-c}}$ for some constant $C > 0$.

We prove this result by induction on s and for each s the proof is by induction on n. Note that for $n = 1$ Eq. (3) holds for all s. Consider now the case $s = 1$. We want to prove that

$$\mathrm{E}F_n(1) = \frac{c}{2-c}\,(n + \theta_{n,1})\;.$$

For the inductive step we use Eq. (1) and get

$$\mathrm{E}(F_{t+1}(1)) = \mathrm{E}F_t(1)\left(1 - \frac{1-c}{t}\right) + c = \frac{c}{2-c}\,(t+\theta_{t,1})\left(1 - \frac{1-c}{t}\right) + c$$

$$= \frac{c}{2-c}\left(t + 1 + \theta_{t,1}\left(1 - \frac{1-c}{t}\right)\right)\;.$$

Since

$$C\left(1 - \frac{1-c}{t}\right) \le C,$$

this finishes the proof for $\alpha = 0$ and $s = 1$.

For $s > 1$ we use Eq. (2) and get

$$\mathrm{E}(F_{t+1}(s)) = \mathrm{E}F_t(s)\left(1 - \frac{s(1-c)}{t}\right) + \mathrm{E}F_t(s-1)\frac{(s-1)(1-c)}{t}$$

$$= \frac{c(s-1)!\,\Gamma\left(2+\frac{1}{1-c}\right)(t+\theta_{t,s})}{(2-c)\Gamma\left(s+1+\frac{1}{1-c}\right)}\left(1 - \frac{s(1-c)}{t}\right)$$

$$+ \frac{c(s-1)!\,\Gamma\left(2+\frac{1}{1-c}\right)(1-c)(t+\theta_{t,s-1})}{(2-c)\Gamma\left(s+\frac{1}{1-c}\right)t}$$

$$= \frac{c(s-1)!\,\Gamma\left(2+\frac{1}{1-c}\right)}{(2-c)\Gamma\left(s+1+\frac{1}{1-c}\right)}\left(t+1+\theta_{t,s}\left(1 - \frac{s(1-c)}{t}\right) + \theta_{t,s-1}\frac{s(1-c)+1}{t}\right)\;.$$

To finish the proof we need to show that

$$(s-1)^{\frac{1}{1-c}}\frac{s(1-c)+1}{t} \le s^{\frac{1}{1-c}}\frac{s(1-c)}{t}\;.$$

It is easy to show that the above inequality holds.

Now we consider the case $p(n) = cn^{-\alpha}$ for $0 < \alpha \le 1$. Let us prove that in this case

$$\mathrm{E}F_n(s) = \frac{c(s-1)!\,\Gamma(3-\alpha)}{(2-\alpha)\Gamma(s+2-\alpha)}\left(n^{1-\alpha} + \theta_{n,s}\right),$$

where $\theta_{n,s} \le Cn^{\max\{0,1-2\alpha\}}s^{1-\alpha+\epsilon}$ for some constant $C > 0$ and for any $\epsilon > 0$.

The proof is similar to the case $\alpha = 0$. Again, for $n = 1$ the theorem holds. Consider $s = 1$. We want to prove that

$$\mathrm{E}F_n(1) = \frac{c}{2-\alpha}\left(n^{1-\alpha} + \theta_{n,1}\right).$$

Inductive step in this case becomes

$$\mathrm{E}(F_{t+1}(1)) = \mathrm{E}F_t(1)\left(1 - \frac{1 - ct^{-\alpha}}{t}\right) + ct^{-\alpha}$$

$$= \frac{c}{2-\alpha}\left(t^{1-\alpha} + \theta_{t,1}\right)\left(1 - \frac{1 - ct^{-\alpha}}{t}\right) + ct^{-\alpha}$$

$$= \frac{c}{2-\alpha}\left(t^{1-\alpha} - t^{-\alpha} + ct^{-2\alpha} + (2-\alpha)t^{-\alpha} + \theta_{t,1}\left(1 - \frac{1 - ct^{-\alpha}}{t}\right)\right)$$

$$= \frac{c}{2-\alpha}\left((t+1)^{1-\alpha} + O\left(t^{-\alpha-1}\right) + ct^{-2\alpha} + \theta_{t,1}\left(1 - \frac{1 - ct^{-\alpha}}{t}\right)\right).$$

In order to finish the proof for the case $s = 1$ it is sufficient to show that

$$O\left(t^{-\alpha-1}\right) + ct^{-2\alpha} \le Ct^{\max\{0,1-2\alpha\}}\frac{1 - ct^{-\alpha}}{t},$$

which holds for sufficiently large C.

For $s > 1$ we have:

$$\mathrm{E}(F_{t+1}(s)) = \mathrm{E}F_t(s)\left(1 - \frac{s\left(1 - ct^{-\alpha}\right)}{t}\right) + \mathrm{E}F_t(s-1)\frac{(s-1)\left(1 - ct^{-\alpha}\right)}{t}$$

$$= \frac{c(s-1)!\,\Gamma(3-\alpha)}{(2-\alpha)\Gamma(s+2-\alpha)}\left(t^{1-\alpha} + \theta_{t,s}\right)\left(1 - \frac{s\left(1 - ct^{-\alpha}\right)}{t}\right)$$

$$+ \frac{c(s-2)!\,\Gamma(3-\alpha)}{(2-\alpha)\Gamma(s+1-\alpha)}\left(t^{1-\alpha} + \theta_{t,s-1}\right)\frac{(s-1)\left(1 - ct^{-\alpha}\right)}{t}$$

$$= \frac{c(s-1)!\,\Gamma(3-\alpha)}{(2-\alpha)\Gamma(s+2-\alpha)}\left((t+1)^{1-\alpha} + O\left(t^{-\alpha-1}\right) - c(1-\alpha)t^{-2\alpha}\right.$$

$$\left. + \theta_{t,s}\left(1 - \frac{s\left(1 - ct^{-\alpha}\right)}{t}\right) + \theta_{t,s-1}\frac{(s+1-\alpha)\left(1 - ct^{-\alpha}\right)}{t}\right).$$

In order to finish the proof, it remains to show that

$$O\left(t^{-\alpha-1}\right) + c(1-\alpha)t^{-2\alpha} + Ct^{\max\{0,1-2\alpha\}}(s-1)^{1-\alpha+\epsilon}\frac{(s+1-\alpha)\left(1 - ct^{-\alpha}\right)}{t}$$

$$\le Ct^{\max\{0,1-2\alpha\}}s^{1-\alpha+\epsilon}\frac{s\left(1 - ct^{-\alpha}\right)}{t},$$

$$O\left(t^{-\alpha}\right) + \frac{t^{1-2\alpha}c(1-\alpha)}{1-ct^{-\alpha}} \leq Ct^{\max\{0,1-2\alpha\}}\left(s^{2-\alpha+\epsilon} - (s+1-\alpha)(s-1)^{1-\alpha+\epsilon}\right),$$

$$O\left(t^{-\alpha}\right) + \frac{t^{1-2\alpha}c(1-\alpha)}{1-ct^{-\alpha}} \leq Ct^{\max\{0,1-2\alpha\}}s^{1-\alpha+\epsilon}\epsilon,$$

which holds for sufficiently large C.

References

1. http://konect.uni-koblenz.de/plots/bidd
2. Aiello, W., Bonato, A., Cooper, C., Janssen, J., Prałat, P.: A spatial web graph model with local influence regions. Internet Math. **5**(1–2), 175–196 (2008)
3. Arenas, A., Danon, L., Diaz-Guilera, A., Gleiser, P.M., Guimera, R.: Community analysis in social networks. Eur. Phys. J. B **38**(2), 373–380 (2004)
4. Artikov, A., Dorodnykh, A., Kashinskaya, Y., Samosvat, E.: Factorization threshold models for scale-free networks generation. Comput. Soc. Netw. **3**(1), 4 (2016)
5. Barabási, A.-L., Albert, R.: Emergence of scaling in random networks. Science **286**(5439), 509–512 (1999)
6. Barthélemy, M.: Crossover from scale-free to spatial networks. EPL (Europhysics Letters) **63**(6), 915 (2003)
7. Barthélemy, M.: Spatial networks. Phys. Rep. **499**(1), 1–101 (2011)
8. Bender, E.A., Canfield, E.R.: The asymptotic number of labeled graphs with given degree sequences. J. Comb. Theory, Ser. A **24**(3), 296–307 (1978)
9. Boccaletti, S., Latora, V., Moreno, Y., Chavez, M., Hwang, D.-U.: Complex networks: structure and dynamics. Phys. Rep. **424**(4), 175–308 (2006)
10. Bollobás, B., Riordan, O., Spencer, J., Tusnády, G., et al.: The degree sequence of a scale-free random graph process. Random Struct. Algorithms **18**(3), 279–290 (2001)
11. Bollobás, B., Riordan, O.M.: Mathematical results on scale-free random graphs. In: Bornholdt, S., Schuster, H.G. (eds.) Handbook of Graphs and Networks: From the Genome to the Internet, pp. 1–34. Wiley-VCH, Weinheim (2003)
12. Bradonjić, M., Hagberg, A., Percus, A.G.: The structure of geographical threshold graphs. Internet Math. **5**(1–2), 113–139 (2008)
13. Buckley, P.G., Osthus, D.: Popularity based random graph models leading to a scale-free degree sequence. Discrete Math. **282**(1), 53–68 (2004)
14. Clauset, A., Newman, M.E.J., Moore, C.: Finding community structure in very large networks. Phys. Rev. E **70**(6), 066111 (2004)
15. da F Costa, L., Rodrigues, F.A., Travieso, G., Villas Boas, P.R.: Characterization of complex networks: a survey of measurements. Adv. Phys. **56**(1), 167–242 (2007)
16. Dunlavy, D.M., Kolda, T.G., Acar, E.: Temporal link prediction using matrix and tensor factorizations. ACM Trans. Knowl. Discovery Data (TKDD) **5**(2), 10 (2011)
17. Ester, M., Kriegel, H.-P., Sander, J., Xu, X., et al.: A density-based algorithm for discovering clusters in large spatial databases with noise. In: KDD, vol. 96, pp. 226–231 (1996)
18. Faloutsos, M., Faloutsos, P., Faloutsos, C.: On power-law relationships of the internet topology. In: ACM SIGCOMM Computer Communication Review, vol. 29, pp. 251–262. ACM (1999)
19. Fortunato, S.: Community detection in graphs. Phys. Rep. **486**(3), 75–174 (2010)

20. Fortunato, S., Barthelemy, M.: Resolution limit in community detection. Proc. Natl. Acad. Sci. **104**(1), 36–41 (2007)
21. Girvan, M., Newman, M.E.J.: Community structure in social and biological networks. Proc. Natl. Acad. Sci. **99**(12), 7821–7826 (2002)
22. Guimera, R., Danon, L., Diaz-Guilera, A., Giralt, F., Arenas, A.: Self-similar community structure in a network of human interactions. Phys. Rev. E **68**(6), 065103 (2003)
23. Holme, P., Kim, B.J.: Growing scale-free networks with tunable clustering. Phys. Rev. E **65**(2), 026107 (2002)
24. Hufnagel, L., Brockmann, D., Geisel, T.: Forecast and control of epidemics in a globalized world. Proc. Natl. Acad. Sci. U.S.A. **101**(42), 15124–15129 (2004)
25. Kempe, D., Kleinberg, J., Tardos, É.: Maximizing the spread of influence through a social network. In: Proceedings of the Ninth ACM SIGKDD International Conference on Knowledge Discovery and Data Mining, pp. 137–146. ACM (2003)
26. Krot, A., Ostroumova Prokhorenkova, L.: Local clustering coefficient in generalized preferential attachment models. In: Gleich, D.F., Komjáthy, J., Litvak, N. (eds.) WAW 2015. LNCS, vol. 9479, pp. 15–28. Springer, Cham (2015). doi:10.1007/978-3-319-26784-5_2
27. Kumpula, J.M., Onnela, J.-P., Saramäki, J., Kertész, J., Kaski, K.: Model of community emergence in weighted social networks. Comput. Phys. Commun. **180**(4), 517–522 (2009)
28. Lancichinetti, A., Fortunato, S., Radicchi, F.: Benchmark graphs for testing community detection algorithms. Phys. Rev. E **78**(4), 046110 (2008)
29. Lipsitch, M., Cohen, T., Cooper, B., Robins, J.M., Ma, S., James, L., Gopalakrishna, G., Chew, S.K., Tan, C.C., Samore, M.H., et al.: Transmission dynamics and control of severe acute respiratory syndrome. Science **300**(5627), 1966–1970 (2003)
30. Lloyd, S.: Least squares quantization in PCM. IEEE Trans. Inform. Theory **28**(2), 129–137 (1982)
31. Masuda, N., Miwa, H., Konno, N.: Geographical threshold graphs with small-world and scale-free properties. Phys. Rev. E **71**(3), 036108 (2005)
32. Menon, A.K., Elkan, C.: Link prediction via matrix factorization. In: Gunopulos, D., Hofmann, T., Malerba, D., Vazirgiannis, M. (eds.) ECML PKDD 2011. LNCS, vol. 6912, pp. 437–452. Springer, Heidelberg (2011). doi:10.1007/978-3-642-23783-6_28
33. Menon, A.K., Elkan, C.: A log-linear model with latent features for dyadic prediction. In: 2010 IEEE 10th International Conference on Data Mining (ICDM), pp. 364–373. IEEE (2010)
34. Miller, K., Jordan, M.I., Griffiths, T.L.: Nonparametric latent feature models for link prediction. In: Advances in Neural Information Processing Systems, pp. 1276–1284 (2009)
35. Newman, M.E.J.: The structure and function of complex networks. SIAM Rev. **45**(2), 167–256 (2003)
36. Newman, M.E.J.: Power laws, pareto distributions and zipf's law. Contemp. Phys. **46**(5), 323–351 (2005)
37. Newman, M.E.J., Girvan, M.: Finding and evaluating community structure in networks. Phys. Rev. E **69**(2), 026113 (2004)
38. Ostroumova Prokhorenkova, L., Samosvat, E.: Recency-based preferential attachment models. J. Complex Netw. **4**(4), 475–499 (2016)

39. Palla, G., Derényi, I., Farkas, I., Vicsek, T.: Uncovering the overlapping community structure of complex networks in nature and society. Nature **435**(7043), 814–818 (2005)
40. Pollner, P., Palla, G., Vicsek, T.: Preferential attachment of communities: the same principle, but a higher level. EPL (Europhysics Letters) **73**(3), 478 (2005)
41. Raigorodskii, A.M.: Small subgraphs in preferential attachment networks. Optimization Lett. **11**(2), 249–257 (2017)
42. Romero, D.M., Meeder, B., Kleinberg, J.: Differences in the mechanics of information diffusion across topics: idioms, political hashtags, and complex contagion on twitter. In: Proceedings of the 20th International Conference on World Wide Web, pp. 695–704. ACM (2011)
43. Wang, C., Knight, J.C., Elder, M.C.: On computer viral infection and the effect of immunization. In: 16th Annual Conference on Computer Security Applications, ACSAC 2000, pp. 246–256. IEEE (2000)
44. Waxman, B.M.: Routing of multipoint connections. IEEE J. Sel. Areas Commun. **6**(9), 1617–1622 (1988)
45. Zhou, T., Yan, G., Wang, B.-H.: Maximal planar networks with large clustering coefficient and power-law degree distribution. Phys. Rev. E **71**(4), 046141 (2005)

Correlation Between Clustering and Degree in Affiliation Networks

Mindaugas Bloznelis[✉] and Justinas Petuchovas

Faculty of Mathematics and Informatics, Vilnius University,
Naugarduko 24, 03225 Vilnius, Lithuania
Mindaugas.Bloznelis@mif.vu.lt

Abstract. We are interested in the probability that two randomly selected neighbors of a random vertex of degree (at least) k are adjacent. We evaluate this probability for a power law random intersection graph, where each vertex is prescribed a collection of attributes and two vertices are adjacent whenever they share a common attribute. We show that the probability obeys the scaling $k^{-\delta}$ as $k \to +\infty$. Our results are mathematically rigorous. The parameter $0 \le \delta \le 1$ is determined by the tail indices of power law random weights defining the links between vertices and attributes.

Keywords: Clustering coefficient · Degree distribution · Random intersection graph · Complex network

1 Introduction and Results

It looks plausible, that in a social network the chances of two neighbors of a given actor to be adjacent is a decreasing function of actor's degree (the total number of its neighbors). Empirical evidence of this phenomenon has been reported in a number of papers, see, e.g., [7,8,13,15]. Theoretical explanations have been derived in [6,13] with the aid of a hierarchical deterministic network model, and in [2] with the aid of a random intersection graph model of an affiliation network. We note that theoretical results [2,6,13] only address the scaling k^{-1}, i.e., $\delta = 1$. In particular, they do not explain empirically observed scaling $k^{-\delta}$ with $\delta \approx 0.75$ reported in [15], see also [8]. In the present paper we develop further the approach of [2] and address the range $0 \le \delta < 1$. The development resorts to a more realistic fitness model of an affiliation network that accounts for variable activities of actors and attractiveness of attributes described below.

An affiliation network defines adjacency relations between actors by using an auxiliary set of attributes. Let $V = \{v_1, \ldots, v_n\}$ denote the set of actors (vertices) and $W = \{w_1, \ldots, w_m\}$ denote the set of attributes. Every actor v_i is prescribed a collection of attributes and two actors v_i and v_j are declared adjacent in the network if they share a common attribute. For example, in the film actor network two actors are adjacent if they have played in the same movie, in the collaboration network two scientists are adjacent if they have coauthored

A. Bonato et al. (Eds.): WAW 2017, LNCS 10519, pp. 90–104, 2017.
DOI: 10.1007/978-3-319-67810-8_7

a publication, in the consumer copurchase network two consumers are adjacent if they have purchased similar products.

A convenient model of a large affiliation network is obtained by linking (prescribing) attributes to actors at random [9,10,12]. Furthermore, in order to model the heterogeneity of human activity, we assign every actor v_j a random weight Y_j reflecting its activity. Similarly, a random weight X_i is assigned to an attribute w_i to model its attractiveness. Now w_i is linked to v_j at random and with probability proportional to the attractiveness X_i and activity Y_j. The random affiliation network obtained in this way is called a random intersection graph, see [5].

We assume in what follows that $X_0, X_1, \ldots, X_m, Y_0, Y_1, \ldots, Y_n$ are independent non-negative random variables. Furthermore, each X_i (respectively Y_j) has the same probability distribution denoted P_X (respectively P_Y). Given realized values $X = \{X_i\}_{i=1}^m$ and $Y = \{Y_j\}_{j=1}^n$ we define the random bipartite graph $H_{X,Y}$ with the bipartition $W \cup V$, where links $\{w_i, v_j\}$ are inserted with probabilities $p_{ij} = \min\{1, X_i Y_j / \sqrt{nm}\}$ independently for each $(i, j) \in [m] \times [n]$. The random intersection graph $\mathcal{G} = G(P_X, P_Y, n, m)$ defines the adjacency relation on the vertex set V: vertices $v', v'' \in V$ are declared adjacent (denoted $v' \sim v''$) whenever v' and v'' have a common neighbor in $H_{X,Y}$. Such a neighbor belongs to the set W and it is called a witness of the edge $v' \sim v''$. We note that for $n, m \to +\infty$ satisfying $m/n \to \beta$ for some $\beta > 0$, the random intersection graph $\mathcal{G}$ admits a tunable global clustering coefficient and power law degree distribution [3,4].

Next we introduce network characteristics studied in this paper. Given a finite graph G and integer $k = 2, 3, \ldots$, define the clustering coefficients

$$c_G(k) = \mathbf{P}\big(v_2^* \sim v_3^* \big| v_2^* \sim v_1^*, v_3^* \sim v_1^*, d(v_1^*) = k\big), \tag{1}$$
$$C_G(k) = \mathbf{P}\big(v_2^* \sim v_3^* \big| v_2^* \sim v_1^*, v_3^* \sim v_1^*, d(v_1^*) \geq k\big). \tag{2}$$

Here (v_1^*, v_2^*, v_3^*) is an ordered triple of vertices of G drawn uniformly at random, $d(v)$ denotes the degree of a vertex v. Note that for a *deterministic* graph G, coefficients (1) and (2) are the respective ratios of subgraph counts

$$\frac{\sum_{v:\, d(v)=k} N_\Delta(v)}{\sum_{v:\, d(v)=k} \binom{d(v)}{2}} \quad \text{and} \quad \frac{\sum_{v:\, d(v)\geq k} N_\Delta(v)}{\sum_{v:\, d(v)\geq k} \binom{d(v)}{2}}. \tag{3}$$

Here $N_\Delta(v)$ and $\binom{d(v)}{2}$ are the numbers of triangles and cherries incident to v. Differently, for the *random* graph $\mathcal{G}$ the conditional probabilities (1) and (2) refer to the two sources of randomness: the random sampling of vertices (v_1^*, v_2^*, v_3^*) and the randomly graph generation mechanism. From the fact that the probability distribution of $\mathcal{G}$ is invariant under permutation of its vertices we obtain that

$$c_{\mathcal{G}}(k) = \mathbf{P}\big(v_2 \sim v_3 \big| v_2 \sim v_1, v_3 \sim v_1, d(v_1) = k\big), \tag{4}$$
$$C_{\mathcal{G}}(k) = \mathbf{P}\big(v_2 \sim v_3 \big| v_2 \sim v_1, v_3 \sim v_1, d(v_1) \geq k\big). \tag{5}$$

An argument bearing on the law of large numbers suggests that for large n, m the ratios (3) can be approximated by respective probabilities (4) and (5).

Our Theorem 2 below establishes a first order asymptotics as $n, m \to +\infty$ of the probabilities (4) and (5)

$$c_{\mathcal{G}}(k) = \left(1 + \beta^{1/2} b(k) a^{-1}(k)\right)^{-1} + o(1), \tag{6}$$

$$C_{\mathcal{G}}(k) = \left(1 + \beta^{1/2} B(k) A^{-1}(k)\right)^{-1} + o(1). \tag{7}$$

Here $a(k), b(k)$ and $A(k), B(k)$ are defined in Theorem 2 below. Our Theorem 1 describes the dependence on k of the leading term of (7). Namely, for a power law distributions P_X and P_Y the leading term of (7) obeys the scaling $k^{-\delta}$.

Theorem 1. *Let $\alpha, \gamma > 5$ and $\beta, c_X, c_Y > 0$. Let $m, n \to \infty$. Assume that $m/n \to \beta$. Suppose that as $t \to +\infty$*

$$\mathbf{P}(X > t) = (c_X + o(1))t^{-\alpha}, \qquad \mathbf{P}(Y > t) = (c_Y + o(1))t^{-\gamma}. \tag{8}$$

Then for $\delta = ((\alpha - \gamma - 1) \wedge 1) \vee (-1)$ we have as $k \to +\infty$

$$\frac{B(k)}{A(k)} = (c + o(1))k^{\delta}. \tag{9}$$

The constant $c = c(\alpha, \gamma, \beta, c_X, c_Y) > 0$ admits an explicit expression in terms of $\alpha, \gamma, \beta, c_X, c_Y$.

It follows from (9) that for large n and m the clustering coefficient $C_{\mathcal{G}}(k)$ obeys the scaling $k^{-\delta}$, where $0 \leq \delta \leq 1$. A related result establishing k^{-1} scaling for $c_{\mathcal{G}}(k)$ has been shown in [2] in the case where P_Y is heavy tailed and P_X is degenerate ($P(X_i = c) = 1$ for some $c > 0$).

We note the "phase transition" in the scaling $k^{-\delta}$ at $\alpha = \gamma + 2$: for $\alpha \geq \gamma + 2$ we have $\delta = 1$ and for $\alpha < \gamma + 2$ we have $\delta < 1$. Our explanation of this phenomenon is as follows. Every attribute w_i forms a clique in $\mathcal{G}$ induced by vertices linked to w_i. Given the weight X_i (of w_i), the expected size of the clique is proportional to X_i. Now, for relatively small α (namely, $\alpha < \gamma + 2$) the sequence $X_1, X_2, \ldots, X_m$ contains sufficiently many large weights so that the corresponding large cliques (formed by attributes) have a tangible effect on the probability (2). Indeed, large cliques may increase the value of (2) considerably.

The proof of Theorem 1 uses known results about the tail asymptotics of randomly stopped sums of heavy tailed independent random variables in the case where the random number of summands is heavy tailed [1]. Similar results are likely to be true also for the local probabilities of randomly stopped sums (work in progress). They would extend Theorem 1 to $c_{\mathcal{G}}(k)$ as well.

Before formulating Theorem 2 we introduce some more notation. We denote $a_r = \mathbf{E} X_0^r$, $b_r = \mathbf{E} Y_0^r$. Let $\beta \in (0, +\infty)$. Let Λ_k, $k = 0, 1, 2$ be mixed Poisson random variables with the distributions

$$\mathbf{P}(\Lambda_k = s) = \mathbf{E} e^{-\lambda_k} \lambda_k^s / s!, \qquad s = 0, 1, \ldots.$$

Here $\lambda_0 = Y_1\beta^{1/2}a_1$ and $\lambda_k = X_k\beta^{-1/2}b_1$ for $k = 1, 2$. Furthermore, for $r = 0, 1, 2, \ldots$ and $k = 0, 1, 2$, let $\Lambda_k^{(r)}$ be a non-negative integer valued random variable with the distribution

$$\mathbf{P}(\Lambda_k^{(r)} = s) = \left(\mathbf{E}\lambda_k^r\right)^{-1}\mathbf{E}\left(e^{-\lambda_k}\lambda_k^{s+r}/s!\right), \qquad s = 0, 1, 2, \ldots.$$

Note that $\Lambda_k^{(0)}$ have the same probability distribution as Λ_k. Let τ_i, $i \geq 1$ be random variables with the probability distribution

$$\mathbf{P}(\tau_i = s) = \frac{s+1}{\mathbf{E}\Lambda_1}\mathbf{P}(\Lambda_1 = s+1), \qquad s = 0, 1, 2 \ldots.$$

Assuming that random variables $\{\tau_i, i \geq 1\}$ are independent of $\Lambda_0^{(r)}$ we introduce the random variables

$$d_*^{(r)} = \sum_{j=0}^{\Lambda_0^{(r)}} \tau_j, \qquad r = 0, 1, 2. \tag{10}$$

We denote for short $d_* = d_*^{(0)} = \sum_{j=1}^{\Lambda_0} \tau_j$.

Theorem 2. *Let $m, n \to \infty$. Assume that $m/n \to \beta$ for some $\beta \in (0, +\infty)$. Suppose that $\mathbf{E}X_1^4 < \infty$ and $\mathbf{E}Y_1^4 < \infty$. Then for each integer $k \geq 2$ relations (6) and (7) hold with*

$$a(k) = a_3b_1^3\mathbf{P}\left(d_*^{(1)} + \Lambda_1^{(3)} = k - 2\right), \quad b(k) = a_2^2b_1^2b_2\mathbf{P}\left(d_*^{(2)} + \Lambda_1^{(2)} + \Lambda_2^{(2)} = k - 2\right),$$
$$A(k) = a_3b_1^3\mathbf{P}\left(d_*^{(1)} + \Lambda_1^{(3)} \geq k - 2\right), \quad B(k) = a_2^2b_1^2b_2\mathbf{P}\left(d_*^{(2)} + \Lambda_1^{(2)} + \Lambda_2^{(2)} \geq k - 2\right).$$

Here we assume that random variables $d_*^{(1)}$ and $\Lambda_1^{(3)}$ are independent. Furthermore, we assume that random variables $d_*^{(2)}$, $\Lambda_1^{(2)}$ and $\Lambda_1^{(2)}$ are independent and $\Lambda_2^{(2)}$ has the same distribution as $\Lambda_1^{(2)}$.

2 Proof

We first prove Theorem 2 and then Theorem 1. Before the proof we introduce some notation. We denote $\{1, 2, \ldots, r\} = [r]$ and $(x)_k = x(x-1)\cdots(x-k+1)$. We denote by $\{w_i \to v_j\}$ the event that w_i and v_j are neighbors in the bipartite graph $H = H_{X,Y}$. We denote

$$\mathbb{I}_{ij} = \mathbb{I}_{\{w_i \to v_j\}}, \qquad \lambda_{ij} = \frac{X_iY_j}{\sqrt{mn}}.$$

Let $\mathbf{P}^* = \mathbf{P}_{X_1,Y_1}$ and $\mathbf{P}^{**} = \mathbf{P}_{X_1,X_2,Y_1}$ denote the conditional probabilities given X_1, Y_1 and X_1, X_2, Y_1 respectively. Furthermore, for $i = 1, 2$, we denote by $\mathbf{P}_{X_i}$ and $\mathbf{P}_{Y_i}$ the conditional probabilities given X_i and Y_i respectively.

Proof of Theorem 2. We only prove (6). The proof of (7) is much the same. Introduce events

$$A = \{v_1 \sim v_2, \, v_1 \sim v_3, \, v_2 \sim v_3\}, \quad B = \{v_1 \sim v_2, \, v_1 \sim v_3\}, \quad K = \{d(v_1) = k\}.$$

We derive (6) from the identity

$$\mathbf{P}(v_2 \sim v_3 \mid v_1 \sim v_2, \, v_1 \sim v_3, \, d(v_1) = k) = \frac{\mathbf{P}(A \cap K)}{\mathbf{P}(B \cap K)} \qquad (11)$$

combined with the relations shown below

$$\mathbf{P}(A \cap K) = n^{-2}\beta^{-1/2}a(k) + o(n^{-2}), \qquad (12)$$
$$\mathbf{P}(B \cap K) = n^{-2}\beta^{-1/2}a(k) + n^{-2}b(k) + o(n^{-2}). \qquad (13)$$

Proof of (12) and (13). Introduce the sets of indices

$$\mathcal{C}_1 = [m], \qquad \mathcal{C}_2 = \{(i,j) : i \neq j; \, i,j \in [m]\},$$
$$\mathcal{C}_3 = \{(i,j,k) : i \neq j \neq k \neq i; \, i,j,k \in [m]\}$$

and split

$$B = B_1 \cup B_2, \quad A = B_1 \cup B_3, \quad B_k = \bigcup_{x \in \mathcal{C}_k} B_{k.x}, \quad k = 1,2,3,$$

where

$$B_{1.i} = \{w_i \to v_1, \, w_i \to v_2, \, w_i \to v_3\},$$
$$B_{2.(i,j)} = \{w_i \to v_1, \, w_i \to v_2, \, w_j \to v_1, \, w_j \to v_3\},$$
$$B_{3.(i,j,k)} = \{w_i \to v_1, \, w_i \to v_2, \, w_j \to v_1, \, w_j \to v_3, \, w_k \to v_2, \, w_k \to v_3\}.$$

We write

$$\mathbf{P}(A \cap K) = \mathbf{P}(B_1 \cap K) + \mathbf{P}((B_3 \cap K) \setminus B_1), \qquad (14)$$
$$\mathbf{P}(B \cap K) = \mathbf{P}(B_1 \cap K) + \mathbf{P}(B_2 \cap K) - \mathbf{P}(B_1 \cap B_2 \cap K) \qquad (15)$$

and evaluate $\mathbf{P}(B_k \cap K)$, for $k = 1,2$, using inclusion-exclusion,

$$\sum_{x \in \mathcal{C}_k} \mathbf{P}(B_{k.x} \cap K) - \sum_{\{x,y\} \subset \mathcal{C}_k} \mathbf{P}(B_{k.x} \cap B_{k.y}) \leq \mathbf{P}(B_k \cap K) \leq \sum_{x \in \mathcal{C}_k} \mathbf{P}(B_{k.x} \cap K).$$
$$(16)$$

We show in Lemma 2 below that the quantities

$$R_k := \sum_{\{x,y\} \subset \mathcal{C}_k} \mathbf{P}(B_{k.x} \cap B_{k.y}), \quad k = 1,2, \qquad (17)$$
$$R_3 := \mathbf{P}((B_3 \cap K) \setminus B_1), \qquad R_4 := \mathbf{P}(B_1 \cap B_2 \cap K)$$

are negligibly small. More precisely, we establish the bounds $R_i = O(n^{-3})$, $1 \leq i \leq 4$. Invoking these bounds in (14)–(16) we obtain

$$\mathbf{P}(A \cap K) = \mathbf{P}(B_1 \cap K) + o(n^{-2}) = m\mathbf{P}(B_{1.1} \cap K) + o(n^{-2}), \qquad (18)$$

$$\mathbf{P}(B \cap K) = \mathbf{P}(B_1 \cap K) + \mathbf{P}(B_2 \cap K) + o(n^{-2}) \qquad (19)$$
$$= m\mathbf{P}(B_{1.1} \cap K) + (m)_2\mathbf{P}(B_{2.(1,2)} \cap K) + o(n^{-2}).$$

In the remaining part of the proof we evaluate the probabilities

$$p_1 := \mathbf{P}(B_{1.1} \cap K) \quad \text{and} \quad p_2 := \mathbf{P}(B_{2.(1,2)} \cap K).$$

We shall show that

$$(nm)^{3/2}p_1 = a(k) + o(1) \qquad \text{and} \qquad (nm)^2 p_2 = b(k) + o(1). \qquad (20)$$

Finally, invoking (20) in (18), (19) we obtain (12), (13) thus proving (6).

It remains to prove (20). For convenience we divide the proof into three steps. For this part of the proof we need some more notation. Let d_1^* (respectively d_2^*) denote the number of neighbors of v_1 in $V^* = \{v_4, v_5, \cdots, v_n\}$ witnessed by the attribute w_1 (respectively w_2). Let d_1' (respectively d_2') denote the number of neighbors of v_1 in V^* witnessed by some attributes from $W_1' = \{w_2, w_3, \ldots, w_m\}$ (respectively $W_2' = \{w_3, w_4, \ldots, w_m\}$).

Step 1. We firstly show that

$$p_1 = \mathbf{P}\big(B_{1.1} \cap \{d_1^* + d_1' = k - 2\}\big) + O(n^{-4}), \qquad (21)$$

$$p_2 = \mathbf{P}\big(B_{2.(1,2)} \cap \{d_1^* + d_2^* + d_2' = k - 2\}\big) + O(n^{-5}). \qquad (22)$$

To show (21) we count neighbors of v_1 in V^*. The number of such neighbors is denoted $d^*(v_1)$. We have $d^*(v_1) = d_1^* + d_1' - d_0$, where d_0 is the number of neighbors of v_1 witnessed by w_1 and by some attribute(s) $w_i \in W_1'$ simultaneously. Combining the inequality

$$d_0 \leq \sum_{j=4}^{n} \left(\mathbb{I}_{1j}\mathbb{I}_{11} \sum_{i=2}^{m} \mathbb{I}_{ij}\mathbb{I}_{i1} \right)$$

with Markov's inequality we obtain

$$\mathbf{P}\big(B_{1.1} \cap \{d_0 \geq 1\}\big) \leq \mathbf{E}\mathbb{I}_{B_{1.1}}d_0 \leq (n-3)(m-1)\mathbf{E}\mathbb{I}_{B_{1.1}}\mathbb{I}_{14}\mathbb{I}_{11}\mathbb{I}_{24}\mathbb{I}_{21}.$$

Furthermore, invoking the inequality

$$\mathbf{E}\mathbb{I}_{B_{1.1}}\mathbb{I}_{14}\mathbb{I}_{11}\mathbb{I}_{24}\mathbb{I}_{21} = \mathbf{E}p_{11}p_{12}p_{13}p_{14}p_{21}p_{24} \leq a_2 a_4 b_1^2 b_2^2 (nm)^{-3}$$

we obtain $\mathbf{P}\big(B_{1.1} \cap \{d_0 \geq 1\}\big) = O(n^{-4})$. Now (21) follows from the fact that the event $B_{1.1}$ implies $d(v_1) = d^*(v_1) + 2$.

The proof of (22) is almost the same. We color w_1 red, w_2 green and all $w_i \in W_2'$ we color yellow. Let d_0' denote the number of neighbors of v_1 witnessed by at least two attributes of different colors. Note that the number $d^*(v_1)$ of neighbors of v_1 in V^* satisfies, by inclusion-exclusion,

$$d_1^* + d_2^* + d_2' - 2d_0' \leq d^*(v_1) \leq d_1^* + d_2^* + d_2'. \tag{23}$$

We combine the inequality

$$d_0' \leq \sum_{j=4}^{n} \left(\mathbb{I}_{11}\mathbb{I}_{1j}\mathbb{I}_{21}\mathbb{I}_{2j} + (\mathbb{I}_{11}\mathbb{I}_{1j} + \mathbb{I}_{21}\mathbb{I}_{2j}) \sum_{i=3}^{m} \mathbb{I}_{i1}\mathbb{I}_{ij} \right)$$

with the identity $\mathbb{I}_{B_{2.(1,2)}}\mathbb{I}_{11}\mathbb{I}_{21} = \mathbb{I}_{B_{2.(1,2)}}$ and obtain, by Markov's inequality and symmetry, that

$$\mathbf{P}\left(B_{2.(1,2)} \cap \{d_0' \geq 1\}\right) \leq \mathbf{E}\mathbb{I}_{B_{2.(1,2)}}d_0' \leq (n-3)\mathbf{E}\mathbb{I}_{B_{2.(1,2)}}\left(\mathbb{I}_{14}\mathbb{I}_{24} + 2(m-2)\mathbb{I}_{14}\mathbb{I}_{31}\mathbb{I}_{34}\right).$$

Furthermore, invoking the inequalities

$$\mathbf{E}\mathbb{I}_{B_{2.(1,2)}}\mathbb{I}_{14}\mathbb{I}_{24} = \mathbf{E}p_{11}p_{12}p_{14}p_{21}p_{23}p_{24} \leq a_3^2 b_1^2 b_2^2 (mn)^{-3},$$

$$\mathbf{E}\mathbb{I}_{B_{2.(1,2)}}\mathbb{I}_{14}\mathbb{I}_{31}\mathbb{I}_{34} = \mathbf{E}p_{11}p_{12}p_{14}p_{21}p_{23}p_{31}p_{34} \leq a_2^2 a_3 b_1^2 b_2 b_3 (mn)^{-7/2},$$

we obtain $\mathbf{P}\left(B_{2.(1,2)} \cap \{d_0' \geq 1\}\right) = O(n^{-5})$. Now (22) follows from (23) and the identity $d(v_1) = d^*(v_1) + 2$.

Step 2. We secondly show that

$$(nm)^{3/2}p_1 = b_1^2\mathbf{E}\left(X_1^3 Y_1 \mathbf{P}\left(\Lambda_1 + d_* = k - 2 \,\middle|\, X_1, Y_1\right)\right) + o(1), \tag{24}$$

$$(nm)^2 p_2 = b_1^2\mathbf{E}\left(X_1^2 X_2^2 Y_1 \mathbf{P}\left(\Lambda_1 + \Lambda_2 + d_* = k - 2 \,\middle|\, X_1, X_2, Y_1\right)\right) + o(1). \tag{25}$$

Let us prove (24). We have

$$\mathbf{P}\left(B_{1.1} \cap \{d_1^* + d_1' = k - 2\}\right) = b_1^2\mathbf{E}(p_{11}p_{12}p_{13}\mathbf{P}^*(d_1^* + d_1' = k - 2)) \tag{26}$$

$$= b_1^2\mathbf{E}(\lambda_{11}\lambda_{12}\lambda_{13}\mathbf{P}^*(d_1^* + d_1' = k - 2)) + o((nm)^{-3/2})$$

and

$$(nm)^{3/2}\mathbf{E}(\lambda_{11}\lambda_{12}\lambda_{13}\mathbf{P}^*(d_1^* + d_1' = k - 2)) = \mathbf{E}(X_1^3 Y_1 \mathbf{P}^*(d_1^* + d_1' = k - 2)) \tag{27}$$

$$= \mathbf{E}(X_1^3 Y_1 \mathbf{P}^*(d_* + \Lambda_1 = k - 2)) + o(1).$$

Here (27) follows from Lemma 1, by Lebesgue's dominated convergence theorem. Furthermore, (26) follows from the inequalities

$$\lambda_{11}\lambda_{12}\lambda_{13} \geq p_{11}p_{12}p_{13} \geq \lambda_{11}\lambda_{12}\lambda_{13}\left(1 - \mathbb{I}_{\{\lambda_{11}>1\}} - \mathbb{I}_{\{\lambda_{12}>1\}} - \mathbb{I}_{\{\lambda_{13}>1\}}\right)$$

combined with the simple bound

$$\mathbf{E}\Big(\lambda_{11}\lambda_{12}\lambda_{13}(\mathbb{I}_{\{\lambda_{11}>1\}} + \mathbb{I}_{\{\lambda_{12}>1\}} + \mathbb{I}_{\{\lambda_{13}>1\}})\Big) = o\Big(\mathbf{E}(\lambda_{11}\lambda_{12}\lambda_{13})\Big) = o((nm)^{-3/2}).$$

Note that (21), (26), (27) imply (24).

The proof of (25) is much the same. We have

$$\mathbf{P}\big(B_{2.(1,2)} \cap \{d_1^* + d_2^* + d_2' = k - 2\}\big)$$
$$= \mathbf{E}\big(p_{11}p_{12}p_{21}p_{23}\mathbf{P}^{**}(d_1^* + d_2^* + d_2' = k - 2)\big)$$
$$= \mathbf{E}\big(\lambda_{11}\lambda_{12}\lambda_{21}\lambda_{23}\mathbf{P}^{**}(d_1^* + d_1' = k - 2)\big) + o((nm)^{-2})$$

and

$$(nm)^2\mathbf{E}\big(\lambda_{11}\lambda_{12}\lambda_{21}\lambda_{23}\mathbf{P}^{**}(d_1^* + d_2^* + d_2' = k - 2)\big)$$
$$= b_1^2\mathbf{E}\big(X_1^2 X_2^2 Y_1^2 \mathbf{P}^{**}(d_1^* + d_2^* + d_2' = k - 2)\big)$$
$$= b_1^2\mathbf{E}\big(X_1^2 X_2^2 Y_1^2 \mathbf{P}^{**}(d_* + \Lambda_1 + \Lambda_2 = k - 2)\big) + o(1).$$

Step 3. In this final step we show that

$$\mathbf{E}\big(X_1^3 Y_1 \mathbf{P}^*(d_* + \Lambda_1 = k - 2)\big) = a_3 b_1 \mathbf{P}(d_*^{(1)} + \Lambda_1^{(3)} = k - 2). \qquad (28)$$
$$\mathbf{E}\big(X_1^2 X_2^2 Y_1 \mathbf{P}^{**}(d_* + \Lambda_1 + \Lambda_2 = k - 2)\big) \qquad\qquad\qquad (29)$$
$$= a_2^2 b_2 \mathbf{P}(d_*^{(2)} + \Lambda_1^{(2)} + \Lambda_2^{(2)} = k - 2).$$

In the proof we use the observation that

$$\mathbf{E}\Big(Y_1^r \mathbf{P}_{Y_1}(d_* = s)\Big) = \mathbf{E}\sum_{i\geq 0}\left(Y_1^r \mathbf{P}_{Y_1}(\Lambda_0 = i)\mathbf{P}\left(\sum_{j=0}^{i}\tau_j = s\right)\right)$$
$$= \sum_{i\geq 0}\left(b_r \mathbf{P}(\Lambda_0^{(r)} = i)\mathbf{P}\left(\sum_{j=0}^{i}\tau_j = s\right)\right)$$
$$= b_r \mathbf{P}\big(d_*^{(r)} = s\big).$$

To show (28) we write the quantity on the left in the form

$$\mathbf{E}\left(X_1^3 Y_1 \sum_{s+t=k-2}\mathbf{P}_{Y_1}(d_* = s) \cdot \mathbf{P}_{X_1}(\Lambda_1 = t)\right).$$
$$= \sum_{s+t=k-2} b_1 \mathbf{P}(d_*^{(1)} = s) \cdot a_3 \mathbf{P}(\Lambda_1^{(3)} = t)$$
$$= b_1 a_3 \mathbf{P}\big(d_*^{(1)} + \Lambda_1^{(3)} = k - 2\big).$$

To show (29) we write the quantity on the left in the form

$$\mathbf{E}\left(X_1^2 X_2^2 Y_1^2 \sum_{s+t+u=k-2} \mathbf{P}_{Y_1}(d_* = s) \cdot \mathbf{P}_{X_1}(\Lambda_1 = t) \cdot \mathbf{P}_{X_2}(\Lambda_2 = u)\right)$$

$$= \sum_{s+t+u=k-2} b_2 \mathbf{P}(d_*^{(2)} = s) \cdot a_2 \mathbf{P}(\Lambda_1^{(2)} = t) \cdot a_2 \mathbf{P}(\Lambda_2^{(2)} = u)$$

$$= b_2 a_2^2 \mathbf{P}\left(d_*^{(2)} + \Lambda_1^{(2)} + \Lambda_2^{(2)} = k - 2\right).$$

□

Proof of Theorem 1. In the proof we use shorthand notation $\tilde{A}(k) = \mathbf{P}(d_*^{(1)} + \Lambda_1^{(3)} \geq k)$ and $\tilde{B}(k) = \mathbf{P}(d_*^{(2)} + \Lambda_1^{(2)} + \Lambda_2^{(2)} \geq k)$. Given two positive functions $f(t)$ and $g(t)$ we denote $f(t) \simeq g(t)$ whenever $f(t)/g(t) \to 1$ as $t \to +\infty$.

Using asymptotic formulas for the tail probabilities of randomly stopped sums $d_*^{(r)}$ reported in [1], and the formulas for the tail probabilities of $\Lambda_k^{(r)}$ shown in Lemma 3, we obtain

$$\mathbf{P}(d_*^{(1)} \geq k) \simeq c_Y \frac{\gamma}{\gamma - 1} a_2^{\gamma - 1} b_1^{\gamma - 2} k^{1 - \gamma}, \tag{30}$$

$$\mathbf{P}(d_*^{(2)} \geq k) \simeq c_Y \frac{\gamma}{\gamma - 2} a_2^{\gamma - 2} b_1^{\gamma - 2} b_2^{-1} k^{2 - \gamma},$$

$$\mathbf{P}(\Lambda_1^{(r)} \geq k) \simeq c_X \frac{\alpha}{\alpha - r} \beta^{(r - \alpha)/2} a_r^{-1} b_1^{\alpha - r} k^{r - \alpha}, \qquad r = 2, 3.$$

Next we combine these asymptotic formulas with the aid of Lemma 4. We have

$$\tilde{A}(k) \simeq \mathbf{P}(d_*^{(1)} \geq k), \qquad \qquad \text{for} \quad \alpha > \gamma + 2, \tag{31}$$

$$\tilde{A}(k) \simeq \mathbf{P}(d_*^{(1)} \geq k) + \mathbf{P}(\Lambda_1^{(3)} \geq k), \quad \text{for} \quad \alpha = \gamma + 2,$$

$$\tilde{A}(k) \simeq \mathbf{P}(\Lambda_1^{(3)} \geq k), \qquad \qquad \text{for} \quad \alpha < \gamma + 2$$

and

$$\tilde{B}(k) \simeq \mathbf{P}(d_*^{(2)} \geq k), \qquad \qquad \text{for} \quad \alpha > \gamma, \tag{32}$$

$$\tilde{B}(k) \simeq \mathbf{P}(d_*^{(2)} \geq k) + \mathbf{P}(\Lambda_1^{(2)} \geq k) + \mathbf{P}(\Lambda_2^{(2)} \geq k), \quad \text{for} \quad \alpha = \gamma,$$

$$\tilde{B}(k) \simeq \mathbf{P}(\Lambda_1^{(2)} \geq k) + \mathbf{P}(\Lambda_2^{(2)} \geq k), \qquad \qquad \text{for} \quad \alpha < \gamma.$$

Finally, from (30), (31), (32) we derive (9).   □

3   Auxiliary Lemmas

Let $\tilde{d}_1^*$ (respectively $\tilde{d}_2^*$) denote the number vertices in $V^* = \{v_4, v_5, \cdots, v_n\}$ linked to the attribute w_1 (respectively w_2). Let $x_1, x_2, y_1 \geq 0$. For $k = 1, 2$, let $\tilde{d}_k, \tilde{\Lambda}_k$ denote the random variables $\tilde{d}_k^*, \Lambda_k$ conditioned on the event $X_k = x_k$

(to get $\tilde{d}_k$, $\tilde{\Lambda}_k$ we replace X_k by a non-random number x_k in the definition of $\hat{d}_k^*$, Λ_k). Let $\hat{d}_1$, $\hat{d}_2$ and $\hat{d}_*$ denote the random variables d_1', d_2' and d_* conditioned on the event $Y_1 = y_1$ (to get $\hat{d}_1$, $\hat{d}_2$ and $\hat{d}_*$ we replace Y_1 by a non-random number y_1 in the definition of d_1', d_2' and d_*).

Lemma 1. *Let $\beta > 0$. Let $n, m \to +\infty$. Assume that $m/n \to \beta$. Assume that $\mathbf{E}X_i^2 < \infty$ and $\mathbf{E}Y_j < \infty$. For any $x_1, x_2, y_1 \geq 0$ and $s, t, u = 0, 1, 2, \ldots$, we have*

$$\mathbf{P}(\hat{d}_1 = s, \tilde{d}_1 = t) \to \mathbf{P}(\hat{d}_* = s, \tilde{\Lambda}_1 = t) = \mathbf{P}(\hat{d}_* = s)\mathbf{P}(\tilde{\Lambda}_1 = t), \qquad (33)$$

$$\mathbf{P}(\hat{d}_2 = s, \tilde{d}_1 = t, \tilde{d}_2 = u) \to \mathbf{P}(\hat{d}_* = s, \tilde{\Lambda}_1 = t, \tilde{\Lambda}_2 = u) \qquad (34)$$
$$= \mathbf{P}(\hat{d}_* = s)\mathbf{P}(\tilde{\Lambda}_1 = t)\mathbf{P}(\tilde{\Lambda}_2 = u).$$

We remark that (33) tells us that random vector $(\hat{d}_1, \tilde{d}_1)$ converges in distribution to the random vector $(\hat{d}_*, \tilde{\Lambda}_1)$. Similarly, (34) tells us that random vector $(\hat{d}_1, \tilde{d}_1, \tilde{d}_2)$ converges in distribution to the random vector $(\hat{d}_*, \tilde{\Lambda}_1, \tilde{\Lambda}_2)$. In particular, (33) implies for any $r = 0, 1, 2, \ldots$ that

$$\mathbf{P}(\hat{d}_1 + \tilde{d}_1 \geq r) \to \mathbf{P}(\hat{d}_* + \tilde{\Lambda}_1 \geq r) \qquad \text{and} \qquad \mathbf{P}(\hat{d}_1 + \tilde{d}_1 = r) \to \mathbf{P}(\hat{d}_* + \tilde{\Lambda}_1 = r)$$

as $n, m \to +\infty$. (34) implies that

$$\mathbf{P}(\hat{d}_2 + \tilde{d}_1 + \tilde{d}_2 \geq r) \to \mathbf{P}(\hat{d}_* + \tilde{\Lambda}_1 + \tilde{\Lambda}_2 \geq r),$$
$$\mathbf{P}(\hat{d}_2 + \tilde{d}_1 + \tilde{d}_2 = r) \to \mathbf{P}(\hat{d}_* + \tilde{\Lambda}_1 + \tilde{\Lambda}_2 = r).$$

Proof of Lemma 1. Before the proof we introduce some notation. Let $\mathbf{P}_1$ denote the conditional probability given $\{Y_4, Y_5, \ldots, Y_n\}$. For $a > 0$ and $s = 0, 1, 2 \ldots$ we denote by $f_s(a) = a^s e^{-a}/s!$ the Poisson probability. Below we use the fact that $|f_s(a) - f_s(b)| \leq |a - b|$. Furthermore we denote

$$\tilde{\lambda}_k = x_k \beta^{-1/2} b_1 \qquad \text{and} \qquad \tilde{\lambda}_{3|k} = \sum_{j=4}^{n} \tilde{\lambda}_{kj}, \quad \tilde{\lambda}_{4|k} = \sum_{j=4}^{n} \tilde{p}_{kj}, \qquad k = 1, 2.$$

Here $\tilde{p}_{kj}$, $\tilde{\lambda}_{kj}$ are defined in the same way as p_{kj}, λ_{kj}, but with X_k replaced by x_k, for $k = 1, 2$.

Proof of (33). We have

$$\mathbf{P}(\hat{d}_1 = s, \tilde{d}_1 = t) = \mathbf{E}\mathbf{P}_1(\hat{d}_1 = s, \tilde{d}_1 = t) = \mathbf{E}\big(\mathbf{P}_1(\hat{d}_1 = s)\mathbf{P}_1(\tilde{d}_1 = t)\big). \qquad (35)$$

Given $\{Y_4, Y_5, \ldots, Y_n\}$, the random variable $\tilde{d}_1$ is a sum of independent Bernoulli random variables. We invoke Le Cam's inequality, see, e.g., [14],

$$\left|\mathbf{P}_1(\tilde{d}_1 = t) - f_t(\tilde{\lambda}_{4|1})\right| \leq \sum_{j=4}^{n} \tilde{p}_{1j}^2 =: R_1^* \qquad (36)$$

and use simple inequalities

$$\left|f_t(\tilde{\lambda}_{4|1}) - f_t(\tilde{\lambda}_{3|1})\right| \le \left|\tilde{\lambda}_{4|1} - \tilde{\lambda}_{3|1}\right| \le \sum_{j=4}^{n} \tilde{\lambda}_{1j}\mathbb{I}_{\{\tilde{\lambda}_{1j}>1\}} =: R_2^*, \tag{37}$$

$$\left|f_t(\tilde{\lambda}_{3|1}) - f_t(\tilde{\lambda}_1)\right| \le \left|\tilde{\lambda}_{3|1} - \tilde{\lambda}_1\right| = x_1 \left|\sqrt{n/m}\left(n^{-1}\sum_{j=4}^{n} Y_j\right) - \beta^{-1/2}b_1\right|. \tag{38}$$

Note that $\left|f_t(\tilde{\lambda}_{3|1}) - f_t(\tilde{\lambda}_1)\right| \to 0$ almost surely, by the law of large numbers. Furthermore,

$$\mathbf{E}R_2^* = (n-4)(nm)^{-1/2}x_1\mathbf{E}Y_4\mathbb{I}_{\{x_1Y_4>\sqrt{nm}\}} = o(1),$$

because $\mathbf{E}Y_4\mathbb{I}_{\{x_1Y_4>\sqrt{nm}\}} = o(1)$. We similarly show that $\mathbf{E}R_1^* = o(1)$. For any $\varepsilon \in (0,1)$ the inequality $\tilde{p}_{1j}^2 \le \tilde{\lambda}_{1j}\left(\varepsilon + \mathbb{I}_{\{\tilde{\lambda}_{1j}>\varepsilon\}}\right)$ implies

$$\mathbf{E}R_1^* = (n-4)\mathbf{E}\tilde{p}_{i4}^2 \le (n-4)\mathbf{E}\tilde{\lambda}_{14}\left(\varepsilon + \mathbb{I}_{\{\tilde{\lambda}_{14}>\varepsilon\}}\right) \le (n-4)(nm)^{-1/2}\left(x_1b_1\varepsilon + o(1)\right).$$

We obtain the bound $\mathbf{E}R_1^* \le \beta^{-1/2}x_1b_1\varepsilon + o(1)$, which implies $\mathbf{E}R_1^* = o(1)$.

Now it follows from (36)–(38) that

$$\mathbf{E}\left(\mathbf{P}_1(\hat{d}_1 = s)\mathbf{P}_1(\tilde{d}_1 = t)\right) = \mathbf{E}\left(\mathbf{P}_1(\hat{d}_1 = s)f_t(\tilde{\lambda}_1)\right) + o(1) \tag{39}$$
$$= \mathbf{P}(\hat{d}_1 = s)f_t(\tilde{\lambda}_1) + o(1).$$

Next we use the fact that $\mathbf{P}(\hat{d}_1 = s) \to \mathbf{P}(\hat{d}_* = s)$. The proof of this fact repeats literally the proof of statement (ii) of Theorem 1 of [3]. Finally, from (35) and (39) we obtain (33):

$$\mathbf{P}(\hat{d}_1 = s, \tilde{d}_1 = t) = \mathbf{E}\left(\mathbf{P}_1(\hat{d}_1 = s)\mathbf{P}_1(\tilde{d}_1 = t)\right) \to \mathbf{P}(\hat{d}_* = s)f_t(\tilde{\lambda}_1).$$

Proof of (34). It is similar to that of (33). We have

$$\mathbf{P}\left(\hat{d}_2 = s, \tilde{d}_1 = t, \tilde{d}_2 = u\right) = \mathbf{E}\left(\mathbf{P}_1(\hat{d}_2 = s)\mathbf{P}_1(\tilde{d}_1 = t)\mathbf{P}_1(\tilde{d}_2 = u)\right). \tag{40}$$

By the same argument as above (see (36)–(38)), we obtain

$$\mathbf{E}\left(\mathbf{P}_1(\hat{d}_2 = s)\mathbf{P}_1(\tilde{d}_1 = t)\mathbf{P}_1(\tilde{d}_2 = u)\right) = \mathbf{E}\left(\mathbf{P}_1(\hat{d}_2 = s)f_t(\tilde{\lambda}_1)\mathbf{P}_1(\tilde{d}_2 = u)\right) + o(1) \tag{41}$$
$$= \mathbf{E}\left(\mathbf{P}_1(\hat{d}_2 = s)f_t(\tilde{\lambda}_1)f_u(\tilde{\lambda}_2)\right) + o(1)$$
$$= f_t(\tilde{\lambda}_1)f_u(\tilde{\lambda}_2)\mathbf{P}(\hat{d}_2 = s) + o(1).$$

Finally, we use the fact that $\mathbf{P}(\hat{d}_2 = s) \to \mathbf{P}(\hat{d}_* = s)$. The proof of this fact repeats literally the proof of statement (ii) of Theorem 1 of [3]. Now from (40), (41) we obtain (34). $\qquad\square$

Lemma 2. *The quantities R_i, $1 \le i \le 4$ defined in (17) satisfy $R_i = O(n^{-3})$.*

Proof of Lemma 2. The bound $R_1 = O(n^{-3})$ is obtained from the identity $R_1 = \binom{m}{2}\mathbf{P}(B_{1.1} \cap B_{1.2})$ and inequalities

$$\mathbf{P}(B_{1.1} \cap B_{1.2}) = \mathbf{E}\mathbb{I}_{11}\mathbb{I}_{12}\mathbb{I}_{13}\mathbb{I}_{21}\mathbb{I}_{22}\mathbb{I}_{23}$$
$$\leq \mathbf{E}\lambda_{11}\lambda_{12}\lambda_{13}\lambda_{21}\lambda_{22}\lambda_{23} = a_3^2 b_2^3 (nm)^{-3}.$$

The bound $R_2 = O(n^{-3})$ follows from inequalities

$$R_2 = \sum_{\{(i,j),(i,r)\}\subset\mathcal{C}_2, j\neq r} \mathbf{P}(B_{2.(i,j)} \cap B_{2.(i,r)})$$

$$+ \sum_{\{(i,j),(k,j)\}\subset\mathcal{C}_2, i\neq k} \mathbf{P}(B_{2.(i,j)} \cap B_{2.(k,j)})$$

$$+ \sum_{\{(i,j),(j,i)\}\subset\mathcal{C}_2} \mathbf{P}(B_{2.(i,j)} \cap B_{2.(j,i)})$$

$$+ \sum_{\{(i,j),(k,r)\}\subset\mathcal{C}_2, i\neq j\neq k\neq r\neq i} \mathbf{P}(B_{2.(i,j)} \cap B_{2.(k,r)})$$

$$= 2^{-1}(m)_3\mathbf{P}(B_{2.(1,2)} \cap B_{2.(1,3)}) + 2^{-1}(m)_3\mathbf{P}(B_{2.(1,3)} \cap B_{2.(2,3)})$$
$$+ (m)_2\mathbf{P}(B_{2.(1,2)} \cap B_{2.(2,1)}) + 2^{-1}(m)_4\mathbf{P}(B_{2.(1,2)} \cap B_{2.(3,4)})$$
$$= 2^{-1}(m)_3\mathbf{E}p_{11}p_{12}p_{21}p_{23}p_{31}p_{33} + 2^{-1}(m)_3\mathbf{E}p_{11}p_{12}p_{21}p_{22}p_{31}p_{33}$$
$$+ (m)_2\mathbf{E}p_{11}p_{12}p_{13}p_{21}p_{22}p_{23} + 2^{-1}(m)_4\mathbf{E}p_{11}p_{12}p_{21}p_{23}p_{31}p_{32}p_{41}p_{43}$$
$$\leq (m)_3\frac{a_3^3 b_1 b_2 b_3}{(nm)^3} + (m)_2\frac{a_3^2 b_2^3}{(nm)^3} + 2^{-1}(m)_4\frac{a_2^4 b_2^2 b_4}{(nm)^4}.$$

The bound $R_3 = O(n^{-3})$ is obtained from the inequalities

$$\mathbf{P}(B_3) \leq \sum_{x\in\mathcal{C}_3} \mathbf{P}(B_{3.x}) = (m)_3\mathbf{P}(B_{3.(1,2,3)})$$

$$= (m)_3\mathbf{E}p_{11}p_{12}p_{21}p_{23}p_{32}p_{33} \leq \frac{(m)_3}{(nm)^3}a_2^3 b_2^3.$$

The bound $R_4 = O(n^{-3})$ is obtained from the inequalities

$$\mathbf{P}(B_1 \cap B_2) \leq \sum_{y\in\mathcal{C}_2} \mathbf{P}(B_1 \cap B_{2.y}) = m(m-1)\mathbf{P}(B_1 \cap B_{2.(1,2)}), \quad (42)$$
$$\mathbf{P}(B_1 \cap B_{2.(1,2)}) \leq \mathbf{P}(B_{1.1} \cap B_{2.(1,2)}) + \mathbf{P}(B_{1.2} \cap B_{2.(1,2)})$$
$$+ (m-2)\mathbf{P}(B_{1.3} \cap B_{2.(1,2)})$$

and bounds

$$\mathbf{P}(B_{1.1} \cap B_{2.(1,2)}) = \mathbf{P}(B_{1.2} \cap B_{2.(1,2)}) = \mathbf{E}p_{11}p_{12}p_{13}p_{21}p_{23} \leq a_2 a_3 b_1 b_2^2 (nm)^{-5/2},$$
$$\mathbf{P}(B_{1.3} \cap B_{2.(1,2)}) = \mathbf{E}p_{11}p_{12}p_{21}p_{23}p_{31}p_{32}p_{33} \leq a_2^2 a_3 b_2^2 b_3 (nm)^{-7/2}.$$

$\square$

Lemma 3. *Let* $\alpha, c > 0$. *Let* r *be an integer and* $0 \le r < \alpha$. *Let* $t \to +\infty$. *For a non-negative random variable* Z *satisfying* $\mathbf{P}(Z > t) = (c + o(1))t^{-\alpha}$ *we have*

$$\mathbf{E}\big(Z^r \mathbb{I}_{\{Z>t\}}\big) = (c + o(1))\alpha(\alpha - r)^{-1} t^{r-\alpha}. \tag{43}$$

Denote $h_r = \mathbf{E}Z^r$. *For a random variable* Λ_Z *with the distribution* $\mathbf{P}(\Lambda_Z^{(r)} = k) = h_r^{-1}\mathbf{E}(e^{-Z}Z^{k+r}/k!)$, $k = 0, 1, 2, \ldots$, *we have*

$$\mathbf{P}(\Lambda_Z^{(r)} > t) = (1+o(1))h_r^{-1}\mathbf{E}\big(Z^r\mathbb{I}_{\{Z>t\}}\big) = (1+o(1))h_r^{-1}c\alpha(\alpha-r)^{-1}t^{r-\alpha}. \tag{44}$$

Proof of Lemma 3. Denote $F(x) = \mathbf{P}(Z \le x) = 1 - \bar{F}(x)$. To show (43) for $r = 1, 2, \ldots$ we apply integration by parts formula for the Lebesgue-Stieltjes integral

$$\mathbf{E}(Z^r\mathbb{I}_{\{Z>t\}}) = \int_t^{+\infty} x^r dF(x) = -\int_t^{+\infty} x^r d\bar{F}(x)$$

$$= t^r \bar{F}(t) + \int_t^{+\infty} rx^{r-1}\bar{F}(x)dx$$

and invoke $\bar{F}(x) = \mathbf{P}(Z > x) = (c + o(1))x^{-\alpha}$.

Proof of (44). Fix r. For $s, t, x > 0$ and $k = 0, 1, 2, \ldots$ we denote

$$S_x^{(k)}(s) := \sum_{i<s} e^{-x}x^{i+k}/i!, \qquad \bar{S}_x^{(k)}(t) := \sum_{i\ge t} e^{-x}x^{i+k}/i!$$

For $0 < s < x < t$ we will use the inequalities (see [11])

$$S_x^{(0)}(s) \le e^{s-x}(x/s)^s \qquad \text{and} \qquad \bar{S}_x^{(0)}(t) \le e^{t-x}(x/t)^t. \tag{45}$$

Given $0 < \varepsilon < 1$ we write for short $t_1 = t(1 - \varepsilon)$, $t_2 = t(1 + \varepsilon)$ and split the probability

$$\mathbf{P}(\Lambda_Z^{(r)} > t) = h_r^{-1}\mathbf{E}\bar{S}_Z^{(r)}(t) = h_r^{-1}(I_1 + I_2 + I_3), \qquad I_k = \mathbf{E}\bar{S}_Z^{(r)}(t)\mathbb{I}_{\{Z\in A_k\}},$$
$$A_1 = [0, t_1), \quad A_2 = [t_1, t_2], \quad A_3 = (t_2, +\infty).$$

We let $\varepsilon = t^{-1/3}$ and evaluate I_1, I_2 and I_3. The second inequality of (45) implies

$$I_1 = \mathbf{E}\big(Z^r\bar{S}_Z^{(0)}(t)\mathbb{I}_{\{Z<t_1\}}\big) \le \mathbf{E}\big(e^{-Z}Z^{t+r}(e/t)^t\mathbb{I}_{\{Z<t_1\}}\big) \le e^{-t_1}t_1^{t+r}(e/t)^t. \tag{46}$$

In the last step we used the fact that $z \to e^{-z}z^{t+r}$ is an increasing function on $(0, t_1)$. Furthermore, the quantity on the right of (46) is less than

$$t^r e^{t-t_1}(t_1/t)^t = t^r e^{\varepsilon t}(1 - \varepsilon)^t = t^r e^{t\varepsilon + t\ln(1-\varepsilon)} \le t^r e^{-t\varepsilon^2/2} = o(t^{r-\alpha}).$$

Hence $I_1 = o(t^{r-\alpha})$. While estimating I_2 we use the inequalities $t_1^{-\alpha} - t_2^{-\alpha} \le c'\alpha\varepsilon t^{-\alpha-1}$ and $\bar{S}_x^{(0)}(t) \le 1$. We obtain

$$I_2 \le \mathbf{E}Z^r\mathbb{I}_{\{t_1\le Z\le t_2\}} \le t_2^r\mathbf{P}(t_1 \le Z \le t_2) = t_2^r(t_2^{-\alpha} - t_1^{-\alpha})c(1 + o(1)) = o(t^{r-\alpha}).$$

We finally evaluate I_3. From the identity $S_x^{(0)}(t) + \bar{S}_x^{(0)}(t) = 1$ we obtain $\bar{S}_x^{(r)}(t) = x^r(1 - S_x^{(0)}(t))$. Using this expression we write I_3 in the form

$$I_3 = \mathbf{E}(Z^r \mathbb{I}_{\{Z>t_2\}}) + R, \qquad \text{where} \qquad R = \mathbf{E}(Z^r S_Z^{(0)}(t) \mathbb{I}_{\{Z>t_2\}}).$$

Note that (43) implies

$$\mathbf{E}(Z^r \mathbb{I}_{\{Z>t_2\}}) = (c + o(1))\alpha(\alpha - r)^{-1}t^{r-\alpha}.$$

We complete the proof by showing that $R = o(t^{r-\alpha})$. The first inequality of (45) implies

$$R \leq \mathbf{E}(Z^{t+r}e^{-Z}(e/t)^t \mathbb{I}_{\{Z>t_2\}}) \leq t_2^{t+r}e^{-t_2}(e/t)^t = t_2^r e^{-\varepsilon t}(1+\varepsilon)^t$$
$$\leq t_2^r e^{-\varepsilon^2 t/4} = o(t^{r-\alpha}).$$

In the second inequality we used the fact that the function $z \to z^{t+r}e^{-z}$ decreases on $(t_2, +\infty)$. In the last inequality we estimated $\ln(1 + \varepsilon)^t = t \ln(1 + \varepsilon) \leq t(\varepsilon - \varepsilon^2/4)$. $\qquad\square$

In the next lemma we collect several simple facts used in the proof of Theorem 1.

Lemma 4. *Let $\alpha \geq \beta > 0$ and $a, b > 0$. Let $t \to +\infty$. Let η, ξ be independent non-negative random variables. Assume that*

$$\mathbf{P}(\eta > t) = (a + o(1))t^{-\alpha} \qquad \text{and} \qquad \mathbf{P}(\xi > t) = (b + o(1))t^{-\beta}.$$

Put $c = a + b$ for $\alpha = \beta$, and $c = b$ for $\alpha > \beta$. We have

$$\mathbf{P}(\eta + \xi > t) = (c + o(1))t^{-\beta}.$$

References

1. Aleškevičienė, A., Leipus, R., Šiaulys, J.: Tail behavior of random sums under consistent variation with applications to the compound renewal risk model. Extremes **11**, 261–279 (2008)
2. Bloznelis, M.: Degree and clustering coefficient in sparse random intersection graphs. Ann. Appl. Probab. **23**, 1254–1289 (2013)
3. Bloznelis, M., Damarackas, J.: Degree distribution of an inhomogeneous random intersection graph. Electron. J. Comb. **20**(3), #P3 (2013)
4. Bloznelis, M., Kurauskas, V.: Clustering coefficient of random intersection graphs with infinite degree variance. Internet Math. (2017)
5. Bloznelis, M., Godehardt, E., Jaworski, J., Kurauskas, V., Rybarczyk, K.: Recent progress in complex network analysis: models of random intersection graphs. In: Lausen, B., Krolak-Schwerdt, S., Böhmer, M. (eds.) Data Science, Learning by Latent Structures, and Knowledge Discovery. SCDAKO, pp. 69–78. Springer, Heidelberg (2015). doi:10.1007/978-3-662-44983-7_6
6. Dorogovtsev, S.N., Goltsev, A.V., Mendes, J.F.F.: Pseudofractal scale-free web. Phys. Rev. E. **65**, 066122 (2002)

7. Eckmann, J.-P., Moses, E.: Curvature of co-links uncovers hidden thematic layers in the world wide web. PNAS **99**, 5825–5829 (2002)
8. Foudalis, I., Jain, K., Papadimitriou, C., Sideri, M.: Modeling social networks through user background and behavior. In: Frieze, A., Horn, P., Prałat, P. (eds.) WAW 2011. LNCS, vol. 6732, pp. 85–102. Springer, Heidelberg (2011). doi:10.1007/978-3-642-21286-4_8
9. Godehardt, E., Jaworski, J.: Two models of random intersection graphs and their applications. Electron. Notes Discrete Math. **10**, 129–132 (2001)
10. Karoński, M., Scheinerman, E.R., Singer-Cohen, K.B.: On random intersection graphs: the subgraph problem. Comb. Probab. Comput. **8**, 131–159 (1999)
11. Mitzenmacher, M., Upfal, E.: Probability and Computing. Randomized Algorithms and Probabilistic Analysis. Cambridge University Press, Cambridge (2005)
12. Newman, M.E.J., Strogatz, S.H., Watts, D.J.: Random graphs with arbitrary degree distributions and their applications. Phys. Rev. E **64**, 026118 (2001)
13. Ravasz, L., Barabási, A.L.: Hierarchical organization in complex networks. Phys. Rev. E **67**, 026112 (2003)
14. Steele, J.M.: Le Cam's inequality and Poisson approximations. Am. Math. Mon. **101**, 48–54 (1994)
15. Vázquez, A., Pastor-Satorras, R., Vespignani, A.: Large-scale topological and dynamical properties of internet. Phys. Rev. E **65**, 066130 (2002)

Author Index

Printed in the United States
By Bookmasters